ALSO BY ISAAC ASIMOV

Asimov's Chronology of Science and Discovery
Asimov's New Guide to Science
*Asimov's Biographical Encyclopedia of Science
and Technology*
The History of Physics
Asimov's Guide to the Bible
Asimov's Guide to Shakespeare
Isaac Asimov's Treasury of Humor
Asimov's Annotated Don Juan
Asimov's Annotated Paradise Lost
The Annotated Gulliver's Travels
Asimov's Annotated Gilbert and Sullivan

ISAAC ASIMOV'S GUIDE TO EARTH AND SPACE

*

ISAAC ASIMOV'S
GUIDE TO
EARTH
AND
SPACE

*

ISAAC ASIMOV

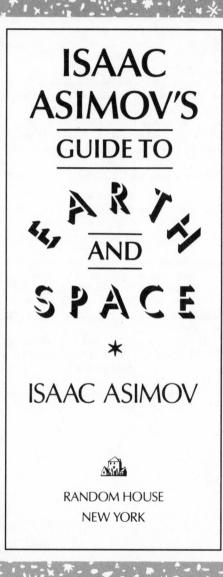

RANDOM HOUSE
NEW YORK

Library of Congress Cataloging-in-Publication Data

Asimov, Isaac
Isaac Asimov's guide to earth & space / by Isaac Asimov.—1st ed.
p. cm.
Includes index.
ISBN 0-679-40437-6
1. Astronomy—Miscellanea. 2. Earth—Miscellanea. I. Title. II. Title:
Isaac Asimov's guide to earth and space.
III. Title: Guide to earth & space.
IV. Title: Guide to earth and space.
QB52.A74 1991   520—dc20   91-11097

Manufactured in the United States of America
24689753
First Edition

Book design by Oksana Kushnir

TO KATE MEDINA
—together again

CONTENTS

ISAAC ASIMOV'S GUIDE TO EARTH AND SPACE

*

The physical world is a large and wonderful place, but it is also confusing, and there is much about it that no one quite understands. There are also many phenomena that some of us understand pretty well, but others do not.

One of the reasons that most of us don't know as much about the world as we might is simply that we don't bother to think about it. Which is not to say that we don't think at all. Everyone thinks, but each person tends to concentrate mostly on matters that seem to be of immediate importance. What shall we have for dinner? How do I pay my bills? Where shall I go for a vacation? How can I go about getting a promotion and a raise in pay? Shall I try to arrange a date with so-and-so? What's this funny pain I have in my side?

These are such important questions to each of us, and our need to answer them is often so strong that there is simply no time to wonder about more general issues such as: What is the shape of the Earth? A natural response to a question like this might be: "Who cares? Why do you bother me with such silly things? What difference does it make?"

But it does make a difference. For example, you can't sail a ship across the ocean and reach your destination by the shortest possible route, or fire a missile and expect it to land on its target, without knowing the shape of the Earth.

But aside from that, and far more important, is that wondering about such questions is fascinating, and finding the answers is fairly easy if you're systematic about it. The aim of this book is to bring these general questions closer to home, by exploring their answers in terms that anyone can follow, making the complexities of the universe absolutely clear.

Of course, one question usually leads to another. Knowledge

about the world is not a straight line but an intricately connected three-dimensional lacework, so that answering a particular question sometimes requires an explanation of something else, which in turn demands the explanation of still another thing, and so on. I will, however, attempt to unravel the threads with as much care as possible so that not too much has to be explained at any one time. Nevertheless, it might still be necessary for me to jump about a bit now and then, and I ask forgiveness for that.

Then, too, as we advance from question to question, simple reasoning in some cases will not be enough; we will have to know a little about what scientists have observed and deduced. But I will try to describe that work with particular care and, wherever possible, without complex mathematics or diagrams. Thinking always leads to more thinking, and there's no end to it. To people who enjoy thinking, that is the glory of science. People who don't enjoy thinking about things that don't concern them immediately, find the necessity of continuing to do so indefinitely frightening, and they turn away from science. I hope you are in the first group.

So let us get started with the question I have already asked and see where that will take us.

1. WHAT IS THE SHAPE OF THE EARTH?

To begin with, we must look around us and see that the Earth is uneven and has no easily described shape. Even if we ignore houses and other man-made objects, and all living things as well, we are still left with an uneven surface of bare rock and soil.

The first conclusion we would come to, then, would be that the Earth is a lumpy object with hills and valleys, cliffs and ravines. In places like Colorado, Peru, or Nepal, where there are towering mountains that reach miles into the air, the Earth's irregularity is very clear. But if you live in some parts of Kansas or Uruguay or the Ukraine, you don't see much in the way of hills or valleys; you see plains, which look pretty flat.

Then, too, even if you do encounter hills and mountains, the Earth may rise on one side, but then fall again on the other side. Valleys and ravines may slant down on one side, but slope up on the other. No part of the Earth's land surface goes up without ever coming down again as you move across it; no part goes down without ever going back up. It seems reasonable, then, to conclude that the Earth is, *on the average*, flat.

Again, if you were to paddle a boat out onto a body of water so large that you couldn't see land in any direction, you would have only the surface of the water to consider. This surface is uneven because it is full of waves. Yet if there is no wind, the waves are not large, and it is easy to see that, on the average, the water surface is flat. In fact, water is much more nearly flat at all times than land.

So it makes sense to suppose that the Earth is flat, and for thousands of years that is exactly what human beings believed. Since a flat Earth made sense and since it didn't take much thinking to see that it made sense, why would anyone waste any further time thinking about it?

Have you ever stood on a hilltop and looked out on the valley below? The valley looks pretty flat and you can look farther and farther outward, past houses, trees, rivers, and other faraway objects, though the more distant they appear to be, the less detail you can make out. What's more, the air isn't usually absolutely clear; bits of fog and smoke obscure the very distant portion, which becomes a kind of bluish haze where the Earth and sky seem to meet.

The meeting place of earth and sky is called the *horizon*, from a Greek word for *boundary*. If you're looking at a flat section of the Earth, the horizon runs evenly from right to left, and such a line is therefore called *horizontal*.

Suppose, however, that you look in another direction at another hill close by. You can't see past the top of the hill to the other side because you can't see around curves. Therefore as you look at the top of the hill, you see only the sky above it, and not the Earth sloping downward beyond it. There is a sharp line that seems quite close to you that marks off the hill against the sky. Thus, if you are looking out over a stretch of land and see a distant misty horizon, you know you are looking over some pretty flat territory, but if you see a sharp nearby horizon, you are looking at a hilltop.

Imagine that you are out on the ocean on the deck of a ship. It is a clear, bright, sunny day, and the sea is calm. The sea air is usually less dusty and misty than land air, so you look off into the distance, and there is the horizon—sharp. The sea meets the sky in a clear horizontal line. You are clearly looking at a hilltop.

How can that be? There are no hills in the ocean, just flat water. The only answer is that the ocean is not flat, but curved, and from your height on the ship deck, you can see out only so far till your eyesight meets the top of the curve, and you can't see beyond it. If you go up to a higher deck, you can see farther out before the curve cut off your view, and if you go to a lower deck, you can see less far out. What's more, if you stand in one place and look all around you, you will see that same sharp horizon at the same distance in every direction; not only does the ocean surface curve, but it curves in the same way and to the same extent in every direction—at least as nearly as one can make out with the eyes.

But why should the ocean curve? It must be following the surface of the Earth, and the Earth itself must be curving in all directions, too. The curving is more obscure on the land because the land is more uneven than the sea, and the air over the land is usually mistier.

Given that the Earth does curve, what kind of curve is it? If the Earth curves in the same way in all directions, it must be a sphere, for that is the only known surface that curves downward equally in all directions. So just by looking and thinking we can see that the Earth is a sphere.

You might ask why people didn't study the horizons and come to this conclusion thousands of years ago, but the trouble is, few people thought about it at all. It was much simpler to think of the Earth as flat, and flatness didn't raise any particular problems in ancient times. A spherical Earth, as we shall soon see, does raise problems that require further thought.

You might ask: Can we trust our eyes? Is looking at the horizon enough? Actually, in this case it is, though we are frequently misled by our eyes, if we don't examine the evidence carefully.

For instance, suppose you are at sea and can identify a ship in the distance sailing toward the horizon. You watch it, and as it approaches the horizon, you don't see the lower decks anymore; then, after a while, you don't the the upper decks, either. All you see are the smokestacks (or the sails, if it is a sailing vessel), and then they disappear, too. It's not just a matter of distance, for if you had a spyglass and watched through that, the ship would seem much larger and closer, but you would still see it disappear first at the bottom, then higher up, then still higher. What you are seeing is the ship sailing over the top of the Earth's curve and down the other side.

The first person we know of who ever maintained that the Earth was a sphere was the Greek philosopher Pythagoras (c. 580–c. 500 B.C.), who came up with this hypothesis in about 500 B.C.

There are other pieces of evidence that show the Earth to be a sphere. Certain stars are visible from some points on Earth and not from others, and during an eclipse of the moon a shadow of the Earth falls on the moon that is always curved like the edge of a sphere. The Greek philosopher Aristotle (384–322 B.C.) listed all the evidence for the Earth's sphericity in about 340 B.C., and though it wasn't commonly accepted at the time, no educated man has doubted it since. In the Space Age of today, photographs have been taken of Earth from outer space in which we can actually *see* that it is a sphere.

2. WHAT IS THE SIZE OF THE EARTH?

As long as people thought the Earth was flat, there was not much reason to worry about how large it was. It might stretch on forever, for all anyone knew, but *forever* is a hard concept to imagine. It was a lot easier to think that the Earth had a definite size and that there was some end to it somewhere. Even today, people speak of "traveling to the ends of the Earth," though nowadays that is only a colorful phrase and not meant to be taken literally.

Of course, the thought of an end to the Earth creates problems. Suppose you traveled a long distance and finally reached the end. Could you fall off? If the ocean stretched out to the end, might it pour off until it was all gone? People who bothered to think of such matters had to work out a way of keeping that from happening. Perhaps the world is rimmed by a solid bank of tall mountains, so that it looks like a frying pan and nothing on the surface can fall off. Or perhaps the sky is a piece of solid matter that is curved like a flattened hemisphere (which is what it looks like) and it comes down to meet the Earth on all sides, so that the Earth is a flat plate with a lid on it—that, too, would keep things in place. Either solution seemed satisfactory.

You might still ask how large the flat world was. In very ancient times, when people could move about only on foot and didn't travel much, it was assumed that the world was quite small and that only one's own region existed. That is why when, in 2800 B.C., there was a tremendous river flood in the Tigris-

Euphrates river valley, the Sumerians who lived there thought the entire world was covered, and that naïve notion has come down to us as the biblical tale of Noah's flood.

As people learned to trade, however, and sent armies hither and yon, and took to riding horses, the horizon of the world expanded, and by 500 B.C., the Persian Empire stretched out from east to west over a distance of 4,800 kilometers (3,000 miles).*

 West of that empire were Greece, Italy, and other lands, and there was no sign of an end.

When the Greek philosophers realized that the Earth was a sphere, however, they knew it had to have a definite size, and you couldn't get away with just saying that it was "very large" or that it went on and on "indefinitely." What's more, the size of the sphere could be judged without necessarily going very far from home.

You see, whereas a flat Earth can stretch out indefinitely, a spherical Earth curves, and the curve must come back on itself. To determine the Earth's size, all you have to do is measure how much it curves; the more sharply it curves, the smaller the sphere, and the more gently it curves, the larger the sphere.

One thing we can be sure of is that the curve is very gentle, so that the sphere is very large. We know this must be so simply because it took so long to decide the Earth was spherical. If the sphere was small, the curvature would be so sharp that it would be impossible not to notice it. The gentler the curvature, the flatter a small region of the Earth would seem.

But how do we measure the extent of the curvature of the Earth?

Here's one way. Take a thin strip of metal and force it down on an absolutely level stretch of the Earth, so that it touches the

*I shall use the metric system in this book and give distances in kilometers, but I will place American units in parentheses. One kilometer is equal to just about five eighths of a mile. The metric system is used over the entire world *except* the United States.

Earth at all points. It will then be forced to follow the curve of the Earth. You can then lift that strip of metal and sight along it and see how much it has curved downward. If the strip of metal is 1 kilometer long, then its downward curve should be about 12½ centimeters (5 inches—for 1 centimeter, which is equal to $1/100,000$ of a kilometer, is equal to about $2/5$ of an inch).

The trouble with taking this measurement is that it would be hard to find a kilometer of Earth's land surface that was absolutely even and to make a strip of metal that followed the curve exactly, so you would end up with a figure you could by no means trust. Even a small error in the shape of the metal strip would cause a big error in calculating the size of the Earth. In other words, some experiments that work perfectly well in theory hardly work at all in practical reality, and this would be one of them. We have to look for something else.

Suppose, then, you have a long, straight rod that is pounded into the Earth partway, so that it stands perfectly erect. If it is a clear, sunny day, and the sun is directly overhead, the rod casts no shadow because the sunlight is coming down about it on every side. But suppose another rod is pounded into the Earth at an angle to the vertical. Now the sunlight strikes the rod and casts a shadow. If there are a series of rods all sticking out of the ground for six feet, but at various angles, each will cast a shadow of a different length. The greater the angle of tipping, the longer the shadow.

Thus, if we measure the lengths of the shadows compared with the length of the rods, we can calculate the angle of tipping without actually measuring the angle. The branch of mathematics that makes this possible is called *trigonometry,* and it was worked out by the ancient Greek mathematicians quite early on. The Greek philosopher Thales (c. 624–c. 546 B.C.) is supposed to have used trigonometry as early as 580 B.C. to measure the height of Egyptian pyramids from the lengths of the shadows they cast.

But you don't have to tip the rods yourself. Suppose you have a rod perfectly erect in one place and another rod perfectly erect in another place hundreds of miles away. In between those two places, the Earth curves, so that if you consider one rod erect, the other is at an angle to it, the size of the angle depending on the amount of curvature of the Earth's surface.

About 240 B.C., the Greek philosopher Eratosthenes (c. 276–c. 196 B.C.) tried to make this very observation. He was told that at the Egyptian city of Syene on June 21, the sun was directly overhead at noon, so that an erect rod cast no shadow. On that same day, in the Egyptian city of Alexandria, where Eratosthenes lived, an erect rod did cast a small shadow.

Eratosthenes measured the length of the shadow and compared it with the length of the rod, and from that he was able to tell how far the curvature of the Earth tilted the rod in Alexandria compared with the rod in Syene. He knew the distance from Syene to Alexandria, so if there was that much curvature in that distance, he could figure out how far the curvature would have to extend before it curved back on itself and completed the sphere. He announced that the Earth's sphere was, to use round numbers in our modern measurements, 40,000 kilometers (25,000 miles) around at the Equator, which is its circumference, and 12,800 kilometers (8,000 miles) from side to side, which is its diameter.

He was quite correct, and, remarkably, the discovery was achieved twenty-two centuries ago, without Eratosthenes ever leaving home, just by clever thinking and a simple measurement.

This doesn't mean, by the way, that Eratosthenes' work was fully accepted. Others made similar measurements and got smaller results, and even as late as the time of Christopher Columbus (1451–1506), the general feeling was that the Earth's circumference was only about 29,000 kilometers (18,000 miles long), less than three fourths of the true circumference. Columbus headed westward in 1492 because he thought that Asia was only 4,800 kilometers (3,000 miles) away. Actually, it was 16,000 kilometers (10,000 miles away), and if the American continents hadn't existed and he hadn't been able to end his journey there, he would never have been heard from again.

The matter was not finally settled till 1522, when an expedition begun by the Portuguese explorer Ferdinand Magellan

(c. 1480–1521) became the first to sail completely around the world. Magellan didn't make it to the end, for he was killed en route in the Philippine Islands, but one ship with eighteen men aboard did make it, and that voyage proved that Eratosthenes' measurement was correct.

3. IF THE EARTH IS A SPHERE, WHY DON'T WE SLIDE OFF?

When children are first told that the Earth is a sphere, it seems to puzzle them. The people on the other side of the Earth (in Australia, for instance, if you are living in the United States) must be walking about with their heads down and their feet up, so why don't they fall off the Earth altogether? After all, if you tried to walk on the ceiling, you would fall off.

It's worse than that, actually. Suppose that you were living at the very top of a spherical Earth (as you certainly seem to be, since the Earth curves gently downward in every direction). In that case, you might be safe only as long as you stayed exactly where you were. If you started moving away in any direction, you would begin to slip down an incline. The farther you went, the steeper the incline until you started sliding faster and faster and more and more helplessly and eventually fell off the Earth altogether. If this were true, all the oceans would have long since tumbled off the Earth, and all the air, too. In other words, we come to the seemingly reasonable conclusion that it is impossible to live on a spherical Earth and therefore that the Earth cannot be a sphere.

But since the Earth *is* a sphere, there must be some mistake in our thinking, and it arises from what we mean when we say *down*. If we are standing erect and we want to indicate the direction *down*, we point to our feet. When we do so, we are also pointing to the center of the Earth, which is about 6,350 kilometers (3,950 miles) below our feet. Assuming that *down* always means the center of the Earth, then wherever you are on the Earth's surface, when you stand erect, the soles of your feet are always facing in that direction. Those Australians, who are also standing erect, have the soles of their feet facing the center of the Earth, too, and to them *down* seems to be in the direction of their feet, just as it does to us.

We are pulled downward, as is everything heavy, which means that we are pulled toward the center of the Earth, as is everything else on the planet's surface, regardless of where it is located. Since there is no sensation that the Earth is curving when we travel, and since its surface seems more or less horizontal, and since *down* is always in the direction of our feet when we are standing erect, the Earth *seems* flat, and nothing ever falls off it, which is another reason it took so long to find out it was spherical. The first person to make it plain that everything on Earth is attracted to its center was Aristotle, and the force responsible for this attraction is called *gravity*, from a Latin word meaning *heavy*.

Suppose you have a huge quantity of matter of any shape, and every part of it is pulled by every other part so that all the material is packed as close as possible. When all the parts are compressed as much as possible and can get no closer, they have taken up the shape of a sphere. No other solid shape has all its parts as close together, on the average, as a sphere does, which is why the Earth, attracting everything to its center, is a sphere.

---

4. DOES THE EARTH MOVE?

To most people in ancient times, this must have seemed the silliest question imaginable; how could there be any doubt over the matter? We can see that the Earth simply does *not* move. To ask the question at all must have seemed a sign of insanity.

Then why did people ask it?

One reason is because everything in the sky moves. The sun rises in the east, moves across the sky, and sets in the west. So does the moon. The stars seem to turn in vast circles about the North Star at the center. Stars that are not close to the North Star make circles large enough to intersect the horizon, so they, too, rise in the east and set in the west.

This motion in the sky didn't surprise most people. To them it seemed natural that the Earth remained absolutely still and motionless and the objects in the sky turned about it, making one turn every day. That was the way it looked, and why should anyone doubt the evidence of the senses? But some people were bound to wonder whether it wasn't possible that the sky was standing still and the Earth was turning under it. To most, that didn't seem a reasonable alternative. It was simply too obvious that the huge Earth was not moving.

But have you ever been on a train, with another train just next to yours, when suddenly the train next to yours started moving slowly backward? You may have been astonished. Why should it be moving backward? You continue watching and, finally, it moves so far backward that the front of that train moves behind your window and you are looking out at the scenery and, behold, the scenery is moving backward, too! You understand at once that it was your train that was moving forward and the other one that was standing still. As long as your train's motion was very smooth, you couldn't tell the difference; you couldn't tell which train was moving and which was standing still.

The ancients didn't have our advantage, however; they weren't accustomed to travel being so smooth that they couldn't tell if they were moving. Walking, running, being pulled in a springless wagon over rutted roads, or riding on a trotting or galloping horse, all produce such irregularities of motion that there could never be any question as to whether you were moving or not. Therefore since the Earth didn't feel as though it were moving, the conclusion was that it was simply not moving.

Now suppose we imagine ourselves back on our train, watching the train next to us moving slowly backward. To check whether it was moving or you were, all you would have to do would be to look in the other direction. Out the window on the other side of the train you would see the station or a town street. If it was moving backward, too, then you would know that it was you that were moving, not the other train. In the case of the Earth and the sky, however, there is nothing neutral to look at.

The first person we know of who suggested that it might be the Earth that was turning and not the sky was the Greek philosopher Heracleides (c. 390–c. 322 B.C.) in about 350 B.C. He wasn't

taken seriously. In 1609, however, an Italian scientist, Galileo Galilei (1564–1642), turned a very primitive telescope on the sky. Among his discoveries was the fact that there were dark spots on the sun. As he watched them from day to day, he noticed that the spots slowly moved about the sun and concluded that the sun was slowly turning, or rotating, about an imaginary line called its axis, making one full turn in nearly twenty-seven days.

If the sun rotated, he thought, why shouldn't the Earth rotate, too, once every twenty-four hours? There was still strong opposition to the notion, and in 1633, the Catholic Church forced Galileo to renounce his views publicly and say that the Earth was motionless.

But that didn't help the conservatives, and in 1665, the Italian-French astronomer Gian Domenico Cassini (1625–1712) was able to show that the planet Mars was rotating every 24½ hours. In 1668, he showed that the planet Jupiter was rotating every 10 hours. After that, scientists began to suspect that the Earth was rotating, too; it just did so in such a regular and smooth fashion that no one could feel it. What's more, the rotation of the Earth did not depend just on the fact that other worlds rotated; there was additional evidence. As astronomers came to realize how large the universe really is (and we'll get to that ourselves later), it became more and more nonsensical to suppose that the Earth was motionless and that the entire vast universe turned around it.

It wasn't until 1851, though, that someone actually demonstrated the rotation so people could actually *see* that it happened. A French physicist, Jean B. L. Foucault (1819–1868), let a long, heavy pendulum swing from the ceiling of a church. It had a spike at the bottom, one that just made a furrow in the sand on the floor of the church. The pendulum kept swinging in the same plane for hour after hour, but the mark in the sand on the floor kept slowly changing direction as the Earth turned under the pendulum. For the first time, crowds watching the pendulum could actually see the Earth turning. Nowadays, of course, we have placed people on the moon, and from there we can actually see the Earth turn.

5. WHEN YOU JUMP UP, WHY DON'T YOU COME DOWN IN A DIFFERENT PLACE?

When astronomers grew insistent that the Earth turned, back in the 1600s, people who didn't believe that raised objections. If the Earth turned, they said, then a person who jumped straight up in the air would have the Earth turn under him and would come back to Earth a small distance from where he started; if you threw a ball straight up into the air, it would come back an even longer distance from where it started; and if a bird flew away from its nest, it would never find its way back. Since none of these things happened, people argued that the Earth could not be moving.

These objections *seemed* to make sense, and if you had just learned that the Earth was rotating, you might be at a loss as to how to counter them, and it would become necessary to think a bit.

Suppose you are seated in a train on a seat just next to the central aisle, with a friend sitting right across the aisle from you. The train is waiting at the station and, having nothing better to do, you toss a ball across the aisle to your friend, who catches it and tosses it back—you have no trouble doing this. Now suppose the train is not waiting at the station but is racing forward along the smooth, straight tracks at 96 kilometers (60 miles) per hour. You toss the ball to your friend—does the motion of the train affect the ball in the air, so that it doesn't reach your friend but hits someone a couple of seats behind him? No, indeed. The ball would go across the aisle just as if the train were standing still. If you just think about it, your experience with the world should be enough to tell you that what I have said about the ball is so, without your having to try it yourself. (An exercise of this kind, one that can be imagined without actually being performed, is called a "thought experiment.")

Why is it as easy to throw the ball on a speeding train as on a stationary train? Because as the train races along the track, everything inside it is moving at that speed, too—you, your friend, the air between you, and the ball you throw across the aisle. If everything is moving forward at the same rate of speed, then it doesn't matter whether that speed is 96 kilometers per hour or zero kilometers per hour.

The Earth rotates at a speed of about 1,600 kilometers (1,000 miles) an hour at the equator, but you and I and the air and any thrown ball moves at the same speed, so you can play baseball anywhere on the planet without worrying about the Earth's motion.

The ancients, of course, didn't have anything like a train, so Galileo used a different thought experiment. Imagine that you are on a sailing ship, which is moving across the sea, scudding before the wind. You climb to the top of the ship's main mast and drop a marlinespike or some other tool that sailors use. The tool falls, but while it is falling, the ship is moving forward so swiftly that by the time it reaches the level of the ship's deck, the ship may have sailed ahead, leaving the tool to drop into the ocean behind it.

There have, however, been thousands of sailing ships on which sailors have accidentally dropped thousands of tools from the tops of masts, and it is common knowledge that the tools *never* drop into the ocean. They invariably drop to the bottom of the mast. While they are falling, they move forward with the ship.

So this type of argument against the Earth's rotation doesn't work. In fact, nobody has ever advanced a single successful argument against the Earth's turning. It turns!

6. WHAT MAKES THE WIND BLOW?

If the air moves with the Earth as it turns, why is there wind? Wind, after all, is moving air, and perhaps it seems to move because, in actual fact, it is standing still and the Earth is moving under it.

Unfortunately, that hypothesis is wrong.

The Earth is turning west to east, which is why all objects in the sky seem to be moving east to west, just as the train next to us seems to be moving backward when we are moving forward. What's more, it is turning west to east at 1,600 kilometers (1,000 miles) per hour at the equator. At distances north and south of the equator, the Earth moves at lower speeds because points on the Earth's surface to the north and south of the equator are traveling through smaller circles in the same amount of time. (At the North and South poles, there is no movement at all.)

If the air was standing still as the Earth turned, we would experience a steady wind going from east to west at the equator at speeds of about 1,600 kilometers (1,000 miles) per hour. Elsewhere, the air speeds would be lower. None of this occurs, so wind can't be caused primarily by Earth's rotation.

When Columbus sailed across the Atlantic in 1492, he found that there were steady winds (now called the *trade winds*) blowing from the east carrying him along. On his way back he sailed northward till he found steady winds blowing from the west (the *westerlies*) that carried him home. This discovery was important, since until then, Western sailors had thought of winds as erratic forces whose presence, absence, and direction depended entirely on the will of divine beings. After Columbus's voyage it became clear that the winds blow according to rule and that they could be exploited in carrying out ocean trade (which is why they were called trade winds.) What was not known at the time was why the winds behaved in such an orderly way.

The first hint of an answer came in 1686. The English scientist Edmund Halley (1656–1742) pointed out that if all the Earth's atmosphere were heated to the same temperature, the air would rest more or less quietly on Earth's surface and there would be no winds to speak of. However, the sun is hottest in the tropics, and the air there is heated to a higher temperature than the air farther north and south. Heated air expands, becomes lighter, and rises, while the cooler air from the north and south moves in to replace it. It is this cooler air coming in that forms the trade winds.

One would think that the cooler air would come in directly from the north in the region north of the equator and directly from the south in the regions south of the equator, but that is not so. The trade winds north of the equator come in from the northeast, while those south of the equator come in from the southeast.

Halley couldn't explain this phenomenon, but in 1735, a British lawyer, George Hadley (1685–1768), did. Cool air from the north is moving at a slower speed than air at the equator, and as this cool air comes south, it retains this slowness, while losing velocity to the faster motion of the Earth, which is moving from west to east. As a result, the wind seems to blow from the northeast. The same effect of the Earth's motion and slower-moving winds to the south of the equator makes the wind seem to blow from the southeast.

Conversely, when air from the equator is forced northward, it is moving *faster* than the ground to the north and gains upon it, which makes the wind seem to come from the west; this produces the westerlies.

This system was worked out in mathematical detail in 1835 by a French physicist, Gaspard Gustave de Coriolis (1792–1843). Thus, this change in the wind's direction due to the varying rotation speeds of different parts of the Earth is known as the Coriolis effect. This effect can make the air whirl in circles with different degrees of force, creating ordinary storms, hurricanes, and even tornadoes.

Winds are important. They air-condition the Earth, distributing heat so that hot areas are not as hot as they might be otherwise and cold areas are not as cold. They carry off water vapor from the ocean as they heat up and they release the water, as rain, when they cool down, so that the continents receive the fresh water that allows life on land to exist.

If we completely understood the rules governing air movements and wind, we could accurately predict the weather, including hot and cold spells, rain, storms, and so on. The trouble is that the rules are so complicated that, even to this day, weather prediction is imperfect.

In fact, we may never be able to predict the weather flawlessly, because we can never measure the initial conditions precisely enough, and even the tiniest change in these conditions can result in enormous differences in the final consequences. This situation is known as *chaos*, and it is becoming apparent that more and more natural phenomena have chaotic properties that cannot be easily predicted, or even predicted at all. This points up a deficiency in science and a limitation of human knowledge, but if there are indeed limitations to what we can do, it is wise to at least know what the limitations are.

7. WHY IS SUMMER WARMER THAN WINTER?

In the previous section I pointed out that temperatures are higher in the tropics than elsewhere on Earth. That is because the sun shines directly down on the tropics so that they receive solar heat in its most concentrated form. Farther north and south, the sunlight comes in at a slant and is spread over a larger area so that the heat is less concentrated.

Nevertheless, people living in northern lattitudes like, say, the United States or Europe know that the weather gets hotter and colder without anyone having to leave home. It is a good deal warmer in July and August than in January and February. (In the Southern Hemisphere, the situation is reversed.) The easiest explanation would be that the sun is closer to the Earth in summer and therefore burns down on us more hotly—but that isn't true. The sun delivers about the same heat all year long.

It is the position of the sun in the sky that counts. If the sun was always shining down directly on the equator, then at noon at any spot along the equator it would always be exactly overhead. Places north of the equator would always see it in the southern sky at noontime, while places south of the equator would always see it to the north. The farther north you were on the Earth's face, the farther south the noontime sun would be, and the farther south you were, the farther north it would be.

The path taken by the sun, however, is at an angle to the equator. The noonday sun shines directly on the equator on March 20th every year, and on that date, night and day are each twelve hours long all over the Earth. This day is known as the vernal equinox, since *vernal* comes from the Latin word for *spring*, and *equinox* from the Latin word for *equal night*.

From day to day, thereafter, the noonday sun moves northward until, on June 21st, it is shining directly over the Tropic of Cancer, which passes just north of Havana, Cuba, at which time it stops and begins to drift southward again. June 21st is therefore called the *summer solstice* (from a Latin word for "sun stands still").

The sun then begins to move southward till it is directly over the equator on September 23 (the *autumnal equinox*) and contin-

ues in this direction until it is shining directly over the Tropic of Capricorn, which runs just south of Rio de Janeiro, Brazil, on December 21st. On this day (the *winter solstice*) the sun stops drifting southward and begins advancing north again, reaching the equator at March 20th and repeating the same cycle year after year.

People in the Northern Hemisphere watch the noonday sun rise higher and higher in the sky until June 21st, then drop lower and lower till December 21st. The higher the sun, the longer the daytime and the shorter the nighttime. In New York City on June 21st, there are sixteen hours of daytime and eight hours of nighttime. The situation is reversed on December 21st, when you have sixteen hours of nighttime and eight hours of daytime. The disparity between night and day is greater the farther north you go, and in the polar region there is a period around the summer solstice during which the sun does not set for anywhere from one whole day to nearly six months, depending on how close to the North Pole you are. Similarly, there is a period around the winter solstice when the sun does not rise at all for an extended length of time.

In the Southern Hemisphere, everything proceeds in reverse. When the noonday sun is rising in the Northern Hemisphere, it is sinking in the Southern Hemisphere, and vice versa. The summer solstice in the north is the winter solstice in the south, and so on.

Naturally, the higher the sun is in the sky and the longer it stays in the sky, the more heat it delivers, so that the Northern Hemisphere gains more heat by day than it loses by night during the summer solstice. The maximum heat comes in July and August—even though by then, the sun is beginning to sink again—because the gain continues to be greater than the loss during those two months. Similarly, more heat is lost at night than is gained by day at the time of the winter solstice, and the coldest times come in January and February. In the Southern Hemisphere, this is reversed, with July and August being the cold months and January and February the hot ones.

In primitive times, the sinking of the sun was watched with particular alarm, for to those who did not understand the irrevocable nature of the sun's rise and fall, there was always the fear that this time the sun would continue to sink forever and eventually disappear. Therefore the coming of the winter solstice, when the sun began to move northward again, was a time of relief and hilarious celebrations. We still have a successor of that celebration in our Christmas–New Year season.

8. HOW DO WE MEASURE TIME?

If we consider the seasons of the year, as we just have, the question naturally arises as to how we measure time.

There are various aspects of time that are psychological and physiological. Thus, time seems to move more slowly when we are ill than when we are well, when we are in pain that when we are not, when we are sad or bored than when we are happy and busy, when we do something we hate than when we do something we love. No matter how inconsistently time seems to rush or creep by, however, it always moves forward. What's more, we all have the feeling that there is an objective aspect to time, that no matter how it seems to us, regardless of our state of mind and body, time is *really* progressing at a steady and unchanging rate. It is this physical time that we are interested in measuring.

Imagine that you had no devices to tell time, but needed to keep track of it. Surely, the logical way—indeed the only way—to measure its passage would be to find some change that was regular and repetitive and count the number of times it occurred. One such change that would have been noticed quite early on by primitive humanity would be the ceaseless alternation of day and night. The days can be easily counted and referred to, and no one has any trouble understanding the meaning of expressions like "today," "tomorrow," "yesterday," "last night," "three days ago," and "five days from now."

Counting days becomes inconvenient, though, if you want to

deal with long periods of time. It is to easy to lose count. Another change, not so obvious, but still well known in prehistoric times, was the changing shape of the moon (its *phases* from a Greek word meaning *appearance*) from night to night. The moon changes from a thin crescent to a full moon and back to a thin crescent, then does the same thing again and again. (We will consider why this happens later in the book.) A complete cycle of the moon takes 29½ days, which is known as the *lunar month*.

The succession of these months can be counted and used as a calendar, the word *calendar* coming from a Latin word for *proclamation,* because the Roman priests would proclaim the night that a new crescent appeared in the sky to mark the beginning of a new month. This is a *lunar calendar,* the word *lunar* coming from the Latin word for *moon.*

The lunar month averaged 29½ days, with alternate months of 29 and 30 days. Twelve of these months were almost long enough to match the cycle of the seasons from spring through summer, autumn, and winter to spring again.

The cycle of the seasons marks a complete year. The season cycle is a little fuzzier than that of the day-night cycle or the moonphase cycle, but its length comes out, on the average, to 365¼ days.

As it happens, twelve lunar months do not exactly match the cycle of the seasons; twelve cycles of the moon's phases are completed in only 354 days—11 days short of the year. This meant that every once in a while a thirteenth month had to be added to keep the cycle of the months even with the cycle of the seasons. This was important, because the calendar had to tell people when it was time to plant or harvest, when to expect the rains of the dry season, and so on. The Babylonians worked out a system for adding the additional month in certain years of a nineteen-year cycle that would keep the lunar calendar exactly equal with the seasons. That calendar is still used for religious purposes in Judaism.

In ancient Egypt, a significant event was the flooding of the Nile River, which spread new, fertile soil over the fields, something that happened at intervals of just about 365 days. This was so important to the Egyptians that they made no attempt to keep to the changes of the moon, and they made each month 30 days long. Then, after twelve months, they added five additional days and began all over. This was a *solar calendar,* from the Latin word for *sun.*

The Egyptian calendar was adopted by Rome in 44 B.C. The five extra days were spread throughout the year, and every fourth year had 366 days to take into account the fact that the year was 365 ¼ days long. With a few further minor changes, this calendar is the one we use today.

9. HOW DO WE MEASURE TIME INTERVALS LESS THAN A DAY LONG?

There are no regular and repetitive natural changes that force themselves on the attention of human beings that occur in less than a day. Yet from ancient times people have felt the need to refer to fractions of a day.

During the daytime, this can be done by considering the sun's position in the sky. We can speak of dawn, when the sun appears on the eastern horizon; morning, when it is still rising in the sky; noontime, when it is at its highest; afternoon, when it is declining in the sky; sunset, when it disappears at the western horizon; and twilight, when dim light precedes actual night. At night, it's a little more difficult, but people who must work by night (notably people on ships) can get a rough idea of the passage of time by the position of the stars that wheel their way about the sky.

Surely, though, we would want to measure time even more

finely, to note the *exact* position of the sun in the sky. The trouble is that observing the sun to determine its exact position will cause blindness. So early man needed to work out a method for measuring the sun without looking at it, and the solution was simple. After all, the sun casts a shadow, and if you drive a stick into the ground, its shadow will be very long at dawn when the sun is at the eastern horizon and, of course, it will point westward. As the sun crosses the sky, the shadow will shorten and will be at minimum length and point due north (in the Northern Hemisphere) at noontime. After that it will begin to lengthen and face in an eastward direction.

By keeping an eye on the shadow, it is possible to follow the sun meticulously without endangering the eyes. Such sundials may have first been used in Egypt about 3000 B.C. The stick or *gnomon* (from a Greek word meaning *one that knows*—the time, of course) was slanted northward so that the end of the shadow would mark out a semicircle, which could be divided into twelve equal stretches called *hours* (from the Greek word for *time of day*). The ancient Sumerians first used twelve as a common dividing number. The sundial worked well in Egypt, where it is sunny almost all the time during the day and where the length of the day does not vary very much in the course of the year. Farther to the north, however, the variation in the length of the day is greater, and the days are often cloudy, at which times the sundial doesn't work at all.

Of course, people could choose some other steady action that

does not depend on sunlight. For instance, they could keep time by burning a candle made of a standard size and composed of some fixed material. A certain length of the candle would burn in, say, one hour. Or you could have dry sand drift from an upper chamber into a lower one through a narrow opening, and know that when all the sand had fallen it might mark, say, two hours. Such devices could work day and night, cloudy or clear, and would be portable besides.

You could continue keeping time by substituting a new candle as the old one burned out or by turning the sand clock over when all the sand had drifted out of the upper chamber. But, even so, these devices had drawbacks. Different candles were bound to burn at different rates, and even the same candle could burn more or less rapidly, depending on such variables as air currents. As for sand glasses, the sand drifted through the opening faster when there was a greater weight of sand above it than when there was less sand there, so it could be used to measure accurately only the time of complete emptying and nothing less.

Perhaps the best clock the ancients had (*clock* is from a French word meaning *bell*, because the hours were announced by the ringing of bells) was the *clepsydra*, or water clock, in which water dropped through a narrow hole from a top chamber to a bottom one. The earliest water clocks have been traced back to 1400 B.C., and by 100 B.C., they were made more efficient by having a continuous flow of water running into the upper chamber, with an overflow. In this way, the upper chamber always had the same head of water, and the rate of drip did not change with time. Eventually, water clocks were fitted with little floats that supported pointers that rose with the water level in the lower chamber. The pointer thus automatically indicated the number of each hour as it passed.

Water clocks were messy, however, and every once in a while, water was spilled and had to be mopped up. In the Middle Ages, therefore, gravity was used. A heavy weight pulled down a cord that was wound around a drive shaft. The weight, as it was pulled down by gravity, forced the drive shaft to turn, and a pointer attached to it marked off the hours on a dial. The trick was to arrange the workings of this instrument so that the pointer turned at a constant speed around the dial once in twelve hours, or twice in one day. In about 1300, something called an *escapement* was invented. This was a device with teeth that engaged the turning drive shaft and allowed it to move a limited distance. Then it disengaged and another tooth caught it, which helped the drive shaft turn slowly enough and constantly enough to measure a full day.

Even the best of these gravity clocks would tend to gain or

lose at least a quarter of an hour in the course of a day, so they had to be checked against sundials periodically. That was good enough for most purposes, but it was not good enough for scientific experiments, which might depend on the exact interval of time in which something happened.

In 1581, Galileo (only seventeen at the time) was attending services at the Cathedral of Pisa and found himself watching a swinging chandelier that was shifting with air currents, now in a wide arc, now in a small one. It seemed to Galileo that whatever the size of the arc, the chandelier swung back and forth in equal times. He tested this by his pulse beat (which can't really be used as a reliable timer because its speed varies with the state of one's mind and with physical activity). Back home, he experimented by suspending weights from strings and allowing them to swing in small arcs and large ones. In this way he discovered the principle of the *pendulum* (from a Latin word meaning *hanging* or *swinging*).

The pendulum has a motion that can, in principle, be used to make the gears of a clock move with great regularity. Its two shortcomings are that it must be kept swinging back and forth and that its beat is not *completely* regular.

In 1656, a Dutch physicist, Christiaan Huygens (1629–1695), made a pendulum swing between two curved guards that forced it to move in a type of arc called a *cycloid*; its period *was* constant. He also worked out methods for using weights to give the pendulum just enough of a push to keep it swinging indefinitely.

Huygens's *pendulum clock* was the first timepiece accurate enough for scientific purposes. It could measure time to the sixtieth of an hour—that is, to the minute—and for the first time, a timepiece could be given two hands. The minute hand made a complete circuit every time the hour hand advanced one hour. Since then, clocks have been constructed that can measure time accurately to the nearest sixtieth of a minute, or second, and a third hand, the second hand, was added.

Today even tiny fractions of a second can be measured accurately.

10. HOW OLD IS THE EARTH?

Now that we have dealt with the measurement of time, let's ask a question about the Earth that involves time: How old is the Earth?

We know quite certainly that the Earth has been in existence at least 5,000 years, for we have written records dating back to 3000 B.C., when the Sumerians invented writing. We have *artifacts*, that is, human-made objects such as pottery and statuettes that date back earlier still. Until nearly 1800, almost everyone in our Western tradition supposed Earth to be about 6,000 years old. Those who believed this did so entirely as a result of their interpretation of the words of the Bible, which they accepted as divine truth, but this was faith and not scientific evidence.

Naturally, there were some, a very few, who gathered evidence and came to conclusions quite different from those the Bible offered. It seemed to these thinkers that the forces of nature—rain, wind, the battering of waves—were slowly changing the face of the Earth. They thought that such forces could account for much of the present-day appearance of the Earth, but only if they had had a chance to work for a long time—much longer than 6,000 years. One man who thought this, in about 1570, was a French scholar, Bernard Palissy (c. 1510–1589).

Those who accepted a 6,000-year age for the Earth did not deny the existence of change, but attributed it all to the legend of Noah's Flood. Palissy refused to believe that any such worldwide flood could possibly have taken place and suggested that the Earth's appearance was due to slow changes over long periods of time. He was burned at the stake in 1589. It was a bad time for people who thought for themselves.

As late as 1681, an English clergyman, Thomas Burnet (c. 1635–1715), wrote a book that supported the story of the Flood, but then, in 1692, he wrote another book in which he questioned the story of Adam and Eve. That ruined his career.

In 1749, a French naturalist, Georges Louis de Buffon (1707–1788), began writing a long encyclopedia in which he tried to explain the world in naturalistic terms. He estimated that in order for the Earth to reach its present state, it had to be at least 75,000 years old. That got him into trouble, and he was forced to recant, as Galileo had.

However, nothing, in the end, can stop people from thinking. The turning point came in 1795, when a Scottish geologist, James Hutton (1726–1797), wrote a book called *The Theory of the Earth* in which he carefully compiled all the evidence in favor of the notion of gradual changes acting over a long time. Over the next half century or so, scientists came to accept Hutton's view—which was called uniformitarianism—of slow, steady change. This theory does not rule out occasional catastrophic incidents, such as gigantic volcanic eruptions, however.

Scientists then began to consider what changes were taking place in the Earth at the present time and to work out how rapidly those changes were happening. If one assumed that these changes had always taken place at the same rate, one could estimate how long they had been going on in order to leave the Earth in its present condition.

The first man to attempt this was Edmund Halley, who had also been the first to figure out what makes the winds blow. In 1715, he considered the saltiness of the sea and reasoned that the salt had been carried there by the rivers, which dissolved small quantities of salt from the land they traversed. Moreover, he found that water can be evaporated from the sea by the sun's heat, but not salt, so that all rain that falls as a result is fresh water, which, as it feeds the rivers and returns to the sea, brings still more salt to the ocean.

If you imagine the ocean to have been fresh water to begin with and calculate how much salt the rivers bring into it each year, you can work out how long the rivers must have been doing this to make the ocean as salty as it is today. This reasoning sounds good, but it has its uncertainties. First of all, perhaps the ocean didn't start as fresh water, but contained some salt from the very beginning. Also, the total amount of salt entering the ocean from the rivers each year was really not known. In Halley's time, virtually nothing was known of rivers outside of Europe. Then there was the possibility that the total amount of salt being carried into the ocean today might be less, or more, than the amount brought in ages past. Not to mention the fact that there are some processes that remove salt from the ocean. Ordinary evaporation doesn't do so, but sometimes shallow arms of the ocean are pinched off and dried up, leaving behind vast areas that become salt mines.

Halley tried to allow for such irregularities and finally decided that for the oceans to be as salty as they are today, the Earth must be about a billion years old. This was a figure that seemed so impossibly large that no one could take it seriously at that time. It was over 13,000 times as long as Buffon's estimate almost three quarters of a century later, but conditions were easier in Great Britain at the time, and Halley did not get into trouble.

Another way of estimating the Earth's age depended upon sedimentation rates. The rivers, lakes, and oceans of the world laid down mud and sludge, which settled to the ground and were called *sediment* (from a Latin word meaning *settlement*). As further sediment was laid down, the weight of the upper layers compressed the lower layers into *sedimentary rock*. People could estimate the rate at which sedimentation was taking place in the present, and if they assumed that this process had been taking place at the same rate all along, they could calculate how long it took to produce the thicknesses of sedimentary rock that were found in the Earth. The results they gathered made it seem as if the Earth had to be over half a billion years old.

These estimates were of the roughest kind; they were suggestive, but they didn't carry conviction. What was needed was some kind of change that was absolutely regular, that had been taking place on Earth from the very beginning, and that could be easily

measured. No one could think what such a change could possibly be in the time of Halley or Hutton, and when it finally showed up on the scene, a century after Hutton, it was discovered entirely by accident.

11. HOW WAS THE AGE OF THE EARTH FINALLY DETERMINED?

In 1896, a French physicist, Antoine Henri Becquerel (1852–1908), quite by accident (he was looking for something else) discovered that a certain substance containing atoms of the metal uranium gave off radiations that were hitherto unknown. The Polish-French chemist Marie Sklodowska Curie (1867–1934) studied the phenomenon further and, in 1898, concluded that the new radiation was the result of *radioactivity*. Uranium and another type of atom, thorium (which is similar to uranium), were both radioactive, and the British chemist Frederick Soddy (1877–1956) was one of those who showed, in 1914, that as a result of radioactivity, uranium and thorium atoms broke down into somewhat simpler atoms, which in turn decayed into others, until finally, at the end of what is termed the *radioactive chain*, atoms of lead were produced. These lead atoms were not radioactive, so the process of decay finally came to an end.

Working with Soddy was the New Zealander Ernest Rutherford (1871–1937), who showed that every radioactive element had what he called a *half-life*. In other words, a given quantity of any radioactive element lost half of its atoms through breakdown in a certain characteristic length of time, then half of what was left in an additional increment of that interval, then half of what was left—and so on. This meant that you could predict exactly how much of any quantity of uranium or thorium would be left after a given number of years.

As it turned out, both uranium and thorium broke down

14. WHAT IS DENSITY?

It might seem to us that a large object would invariably be more massive than a small one, but we know from experience that that is not true. A large object made out of cork may be less massive (and weigh less, for that matter) than a smaller object made out of lead; some materials simply seem to pack more mass into a given volume than others. The amount of mass in a given volume represents its *density*, so we can say that some objects are denser than others.

A cube of water that is 1 centimeter on each side (1 cubic centimeter) weighs just 1 gram.* (This is not a coincidence; the values of the two units were deliberately chosen so that this would be the case.) Since 1 cubic inch is equal to 16.4 cubic centimeters and an ounce is equal to 28.35 grams, the density of water is not only equal to 1 gram per cubic centimeter but also to 0.58 ounces per cubic inch. The use of grams per cubic centimeter is so much more convenient than ounces per cubic inch that for density, at least, I will use only the metric system.

If we know the mass of any object in grams and also its volume in cubic centimeters, we can divide the mass by the volume and get a figure that will represent its density.

The ancient Greeks discovered how to determine the volume of a sphere from its diameter, and since we know the diameter of the Earth, we can calculate its volume. Once Cavendish had worked out the mass of the Earth, he was able to work out its density by dividing the mass by the volume. It turns out that the density of the Earth is, on the average, 5.518 grams per cubic centimeter, and, therefore, 5.518 times as dense as water.

*A centimeter is equal to $\frac{1}{100,000}$ of a kilometer, and a gram is equal to $\frac{1}{1000}$ of a kilogram.

15. IS THE EARTH HOLLOW?

This question may startle some of you who have never for one moment ever thought that the Earth might be hollow, but there are many people who have believed it over the ages, and the idea has inspired many stories and legends. After all, there are caves in the Earth, though they are only superficial. The deepest cave we know, in the western Pyrenees, is only 1.17 kilometers (0.7 miles) deep, which is quite shallow compared with the 6,350 kilometers distance to the Earth's center. There are always people, however, who have the fantasy of finding a cave that will lead them deep into the Earth's hollow interior.

The notion that the Earth might be hollow dates back to primitive times. In the Greek myths, giants who rebelled against Zeus were chained underground, and it was their writhings that were believed to cause earthquakes. The Greeks' Hades and the Jews' Sheol were both supposedly located underground, and the existence of volcanoes seemed to make it certain that the interior of the Earth was a place of fire and brimstone, fit for tortures.

In the early days of science, some scientists found themselves trying to justify this religious notion of a hollow Earth. In 1665, the German scholar Athanasius Kircher (c. 1601–1680) published the most highly regarded geology book of its time, in which he described the Earth as riddled with caverns and tunnels in which dragons lived. In the early 1800s, an American military man, John Cleve Symmes (1742–1814), presented elaborate theories about the hollow Earth and maintained that there were openings in the north polar regions where people could find their way into the Earth's interior. This idea caught the imagination of people in the way that crackpot notions often do, even today, and after Symmes's time, there was a steady drizzle of science fiction books about travels to the center of the Earth. In 1864, the French writer Jules Verne (1828–1905) published the best of them, *A Journey to the Center of the Earth*, in which he described underground oceans, dinosaurs, and ape-men, and suggested that the opening to the interior was in Iceland. Earlier, Edgar Allan Poe (1809–1849) had written such a story, too, with an opening at the North Pole.

Of course, when the American explorer Robert Edwin Peary (1856–1920) reached the North Pole in 1909, it was quite plain that there were no openings into the interior in the far northern regions. But the stories continued nevertheless. The most popular of the hollow Earth stories was a series written by Edgar Rice Burroughs (1875–1950) about Pellucidar, the name he gave to the underground world. The first of these was published in 1913.

And yet we've known since 1798 that the Earth is not hollow and can't be hollow. As soon as Cavendish measured the mass of the Earth, we knew that its density was 5½ grams per cubic centimeter. (The current figure is 5.518.) The density of the rocks in the Earth's crust is, on the average, 2.8 grams per cubic centimeter. If the Earth were hollow, and the hollow were, presumably, filled with air, then the overall density would be *less* than 2.8 grams per cubic centimeter. The fact that Earth's overall density is 5.518 grams per cubic centimeter tells us that the interior must actually be considerably *denser* than the crustal rocks. The Earth simply can't be hollow. There are many other reasons why it can't be, but its density alone is evidence enough.

16. WHAT IS THE EARTH'S INTERIOR REALLY LIKE?

Since we know that the Earth's crust seems to have a density of 2.8 grams per cubic centimeter and that the density of the Earth as a whole is just over 5.5 grams per cubic centimeter, we can see at once that a portion of the Earth's interior density must be higher than 5.5 to bring it up to that overall figure.

If we think about this fact a bit more, it might occur to us that this is only to be expected. The size of an ordinary ball of matter that we can manipulate in the laboratory is so small that its gravitational effects are negligible. In the case of the Earth, however, there is an enormous gravitational pull tugging at its substance. If we suppose that the Earth is rock all the way through, then the deeper layers of that rock must be crushed under the weight of the outer layers. This weight acts to compress the inner layers, to squeeze all its mass into a smaller volume. The deeper layers would therefore have a naturally higher density than the other layers, and that might solve the problem right there.

But it doesn't. Human beings can put a certain amount of pressure on rock and can calculate how much that pressure compresses it and raises its density. It turns out that all the weight of all the outermost layers of the Earth cannot squeeze the deeper layers into anything close to the density required to achieve the 5.5 grams per cubic centimeter average.

We can only conclude, then, that the Earth is not solid rock, and that in the deep layers of the Earth, there must be other materials that are much denser than rock. But what are these materials, and how can we possibly find out anything about them? I have already said that the deepest natural caves are of little depth. The deepest oil well we have drilled is 9.6 kilometers (6 miles) deep, and that is only about 1/670th of the way to the Earth's center.

Are we completely unable, then, to learn anything about the Earth's center? Actually, we are not. There are occasional earthquakes that trouble the Earth's surface and produce powerful vibrations that travel through the Earth's interior in the form of different types of waves. This wave motion resembles the manner in which waves cross the surface of a pond when it is disturbed or the way sound waves travel through air. In fact, some of the earthquake waves, called *primary waves* or *P waves*, have properties of sound waves, whereas other earthquake waves, the *secondary waves*, or *S waves*, have properties of water waves.

These waves travel through the Earth and can emerge at the surface at a significant distance from the original disturbance. The first simple instrument to study such waves, the seismograph, was invented in 1855 by the Italian physicist Luigi Palmieri (1807–1896). Seismographs were rapidly improved in subsequent years, and in the 1890s, the British engineer John Milne (1850–1913) set up a series of seismographs in various parts of the world. There are now over five hundred very elaborate seismographs distributed over the surface of the planet.

From what the seismographs tell us about where and when earthquake waves reappear, scientists can plot the path they take through the Earth's structure. If the properties of the Earth's substance were the same everywhere, those waves would travel at a fixed speed in a straight line. However, since the Earth's density increases with depth, partially due to pressure and compression, the path taken by the waves curves. From the nature of the curves, scientists can tell how much the Earth's density increases at various depths. At some depths, the waves veer sharply in their paths, which signals a sudden change in density due to a difference in chemical structure, rather than the gradual change caused by compression.

The study of earthquake waves divides the Earth's structure into three main divisions. The outermost layer is called the *crust*, which is made up of the familiar rocks we know. About 32 kilometers (20 miles) below the Earth's surface (on the average) there is a sharp change that was first detected in 1909 by a Croatian geophysicist, Andrija Mohorovičić (1857–1936). It is called the *Mohorovičić discontinuity* or, more simply, the *Moho discontinuity*. Below the Earth's crust is the *mantle*, which is also composed of rock. The mantle rock is denser than the rocks of the crust, however, partly because it is compressed and partly because it is a naturally denser material. But the mantle is not dense enough to account for the high overall density of the Earth.

At a depth of 2,900 kilometers (1,800 miles) below the surface of the Earth, the earthquake waves veer sharply again, a fact first demonstrated in 1914 by the German geologist Beno Gutenberg (1889–1960). This inner region of the Earth, the *core*, is, indeed, dense enough to account for Earth's high overall density. Scientists determined the composition of the core by noticing that P waves travel through it, but S waves do not. The properties of the S waves are such that they cannot travel through the body of a liquid, while the P waves can. From this we can deduce that much of the Earth's core is liquid. (The Earth's crust, its mantle, and its liquid core bear very much the same relation to each other that the shell, the white, and the yolk of an egg do—but this is merely an interesting coincidence and nothing more.)

It remains, though, to decide what the core consists of, for it must be made of substances that are denser and have lower melting points than rock. The most common candidates are the various metals, so we suspect that Earth has a liquid metallic core—but which metal?

Actually, a likely answer had been given long before earthquake data revealed the fine details of the Earth's internal structure. Meteorites occasionally strike the Earth's surface (I'll have more to say about them later in the book), and although most of them are rocky in nature, some 10 percent are metallic. These are always iron, along with the related metal nickel, in a 9 to 1 ratio.

The French geologist Gabriel August Daubrée (1814–1896) therefore suggested, as early as 1886, that the Earth's core might be made up of a nickel-iron mixture. This idea seemed reasonable, and most scientists now assume that the Earth's core is indeed 90 percent iron and 10 percent nickel, though there are currently arguments as to whether oxygen or sulfur, or both, are also present in significant quantities.

17. DO THE CONTINENTS MOVE?

Since we've mentioned earthquakes, it is logical to consider what might cause them, and in order to do so, we should first ask whether the continents move. Of course they move in the sense that they are rotating about the Earth's axis as part of the solid globe, but do they each move relative to one another?

It might seem that the answer is an obvious no. How can the continents move? But even in ancient times, there was a feeling that continents, or parts of them, did move, at least in the sense that they would rise or fall. As long ago as 540 B.C., the Greek philosopher Xenophanes (c. 560–c. 480 B.C.) pointed out that seashells were sometimes found embedded in the rocks of mountain heights and that, therefore, there must have been a time when those mountain heights were underwater. Mountains, he maintained, originally came from material that was at sea level and was somehow thrust upward. He was quite correct in this, but no one took him seriously at the time.

About 1889, an American geologist, Clarence Edward Dutton (1841–1912), revived the notion in a much more sophisticated form. He maintained that the rocks making up the continents were less dense than those making up the sea bottom. The continents, therefore, floated considerably higher than the ocean bottoms, and they rose above the ocean surface. Mountainous regions rested on rock that was even less dense, so that they rose above the general level of the land. Dutton called this phenomenon *isostasy*. But even if the continents and portions of continents did move higher and lower, there seemed to be no indication that they ever moved *sideways*.

Yet the map of the world is suggestive. Once the American continents were discovered and their Atlantic shores were mapped, a curious fact emerged, one which was first pointed out in 1620 by the British philosopher Francis Bacon (1561–1626). If you look at the eastern coast of South America, you will see that it fits remarkably well into the western coast of Africa. It is impossible to look at the map and not begin to wonder if Africa and South America were once a single land mass that split in half, with the halves moving apart.

In 1912, a German geologist, Alfred Lothar Wegener (1880–1930), took up the matter in detail and pointed out that the two continents might have drifted apart by floating through the heavier rocks underlying the oceans. In fact, he theorized that all the continents were once a single landmass, which he called *Pangaea* (from Greek words meaning *all Earth*), that broke up into the separate continents. This phenomenon he called *continental drift*. In a way, he was right, but the suggestion that the continents floated across the ocean bottoms simply wouldn't work. The underlying rock was far too stiff for that, so the idea was dismissed as impossible even as late as 1960.

But a new idea arose. In the 1850s, there were attempts being made to lay a cable across the bottom of the Atlantic Ocean, one that would allow for telegraphic contact between North America and Europe. An American oceanographer, Matthew Fontaine Maury (1806–1873), took soundings of the Atlantic Ocean to try to decide what the best route might be for the cable. In the process, he discovered, in 1854, that the depth of the Atlantic Ocean was less in the middle than on either side. He believed that there was a plateau in mid-ocean, which he named *Telegraph Plateau.*

It was very difficult to take soundings of the ocean floor. One had to lower a weighted cable that would have to be many kilometers long, tell when it hit bottom, pull it back up, and measure the length that had been paid out. It was very tedious work, and uncertain, too, and few spots could be measured properly in the course of a single voyage, so that Maury's work was just a beginning.

In 1872, a British expedition under Charles Wyville Thomson (1830–1882) spent four years at sea, traveled some 125,000 kilometers (78,000 miles), and made 372 deep-sea soundings with a 6.4-kilometer cable. Nothing better was done for another half a century, but the expedition still gave only the barest picture of the ocean floor. But then, during World War I, the technique of *echo sounding* was developed. It used ultrasonic sound waves, too high-pitched to be heard by human ears, which penetrated to the ocean bottom and were reflected back in a matter of minutes. From the time lapse between the emission of the waves and the reception of their reflection, the depth of the ocean could be estimated. A German ship first applied this technique to ocean exploration in 1922, and we came to know the nature of the ocean bottom in detail.

The greatest explorer of the ocean bottom was an American geologist, William Maurice Ewing (1906–1974), who took innumerable measurements and showed in the early 1950s that Telegraph Plateau was not a plateau but a long, rugged mountain chain curving down the middle of the Atlantic Ocean, with some of its highest peaks jutting above the ocean's surface to form islands. In 1956, Ewing showed that this mountain range extended around

Africa into the Indian Ocean and around Antarctica into the Pacific Ocean. It was a world-girdling system, which became known as the *Mid-Oceanic Ridge*. In 1957, Ewing showed that there was a deep fault, the *Great Global Rift*, that ran down the center of the ridge and made it look as though the crust of the Earth was broken into a number of tightly fitting plates. These were called *tectonic plates*, from the Greek word for *carpenter*, because they were fitted together as snugly as if a carpenter had done the job.

Another American geologist, Harry Hammond Hess (1906–1969), thought about the tectonic plates and suggested in 1962 that material from the deeper regions of the Earth welled up through the Great Global Rift in the middle of the Atlantic Ocean and pushed the two plates on either side apart. The plate bearing Africa was pushed eastward, and the plate bearing South America was pushed westward, while the ocean in between broadened, a process called *sea-floor spreading*. This concept was rapidly accepted by other geologists. South America and Africa had, indeed, originally been part of the same landmass, as Wegener had suggested, but they had separated not by floating and drifting but by being pushed apart by force. Wegener had come to the right conclusion, but the mechanism he suspected of causing this phenomenon had been wrong. The new mechanism was correct, however, and all of geology is now interpreted on the basis of *plate tectonics*—the study of the slow movement of the tectonic plates.

18. WHAT CAUSES EARTHQUAKES AND VOLCANOES?

Earthquakes and volcanoes have existed from the oldest times. They have always had the capacity for great destruction and have succeeded in frightening human beings dreadfully, for in a few minutes they can kill many thousands of human beings. The most powerful known volcanic eruption in historic times was one that suddenly destroyed the island

of Thera in the Aegean Sea, north of the island of Crete, in about
1500 B.C. The island had shown no signs of volcanic potential in
the memory of the people living there, but deep inside the Earth,
enough pressure had finally built up to blow the top of the volcano
away. Not only was Thera destroyed, an event that probably gave
rise to the legend of Atlantis, but the island of Crete was so badly
damaged by rains of ash and tidal waves that its flourishing civilization did not survive much longer. In fact, the whole eastern
Mediterranean fell into chaos, and the Egyptian Empire went into
permanent decline.

A volcano in Italy near the city of Naples had also been quiet
for so long that people had forgotten its danger. But then, in 79,
it erupted and buried the cities of Pompeii and Herculaneum. The
great Roman writer Pliny (23–79) died because he approached too
closely in his attempt to observe and describe the eruption. It
killed about 4,000 people, but Mt. Etna in Sicily, the tallest and
most continually active of the European volcanoes, killed up to
20,000 in an eruption in 1669. Another huge eruption took place
in Iceland in 1783. In 1815, the Indonesian volcano Tambora blew
up on Sumbawa Island, and in 1883, another Indonesian volcano,
Krakatoa, did the same. In all three cases, there was heavy loss of
life. In 1902, Mt. Pelée, on the West Indian island of Martinique,
erupted, and red-hot, poisonous gases rolled down the side of the
mountain into Martinique's former capital, Saint Pierre. In three
minutes, all but one of the city's 38,000 people were killed. (The
exception was a convicted murderer awaiting execution in an
underground prison.)

Earthquakes can be even more deadly. On January 24, 1556,
an earthquake struck in the province of Shensi in China that is
supposed to have killed 800,000 people in just a few minutes. In
1703, an earthquake killed 200,000 people in Tokyo, and in 1737,
another killed 300,000 in Calcutta. The greatest European earthquake in modern times took place on November 1, 1755, when an
earthquake (followed by a tidal wave and a fire) destroyed the city
of Lisbon, Portugal, and killed 60,000 people. There was also a
terrible earthquake in 1812 along the Mississippi River, near the

present-day town of New Madrid, but so few people lived there at the time that it killed no one.

What causes these phenomena? We can dismiss early theories of vengeful gods, fire spirits, and so on. Aristotle thought that air was trapped in various places underground and that earthquakes were caused by occasional discharges of the air that escaped. But as people began to consider the matter of volcanoes and earthquakes, they noticed that the large majority of them took place in certain areas. Of the five hundred active volcanoes on Earth, nearly three hundred are to be found in a big curve all around the borders of the Pacific Ocean, and about eight more are along the chain of Indonesian islands, with a somewhat lesser number along the line of the Mediterranean Sea. Earthquakes happen most often in these regions as well, which suggested that earthquakes and volcanoes were somehow connected and that the same causes might be responsible for both.

The earthquake at Lisbon prompted a wave of scientific research into the problem and led to the setting up of seismographs here and there, as I mentioned earlier. Then, in 1906, an earthquake destroyed the city of San Francisco, and an American geologist, Harry Fielding Reid (1859–1944), who had come to investigate the site, noticed that the ground near the city had slipped. One side of what looked like a crack in the ground had moved forward compared with the other side. Most people assumed that the crack had been made by the earthquake, but Reid had a different idea: it seemed to him that the crack, or fault, might have been there all the time. (We now call it the San Andreas Fault.) With time, pressure built up to make the two sides of the fault move against each other. Ordinarily the two sides were held in place by friction, but as the pressure built up, one side advanced, rubbing against the other in a series of jerks that caused vibrations strong enough to raze a city and snuff out thousands of lives.

Although Reid was on the right track, the cause of earthquakes didn't become entirely clear until the tectonic plates were discovered. It was then understood that, as a result of forces deep within the Earth, the plates were always moving very slowly, but at the boundaries between plates the forces sometimes induced a sidewise motion such as Reid had noticed in connection with the San Francisco earthquake. The San Andreas Fault is, after all, a section of the boundary between the plate underlying North America and the plate underlying the Pacific Ocean.

Furthermore, at the fault lines around the world there are weak points where hot rock from below can ooze upward and produce volcanic eruptions, and when two plates meet head-on, the edges sometimes crumple, producing mountain ranges. The Himalayan Mountains, the largest upthrusted region on Earth right now, was produced when the plate bearing India plowed slowly into the plate bearing the rest of Asia. Sometimes one plate slides under another, sucking the sea bottom downward and producing the ocean deeps, which in some places can be as deep as eleven kilometers (about seven miles).

It had proven impossible to completely understand the various phenomena involving the Earth's crust until plate tectonics was discovered. It made the seemingly impossible simple—which is always the sign of a good theory.

19. WHAT IS HEAT?

It would now seem reasonable to ask what these forces are, deep in the Earth, that power earthquakes and volcanoes. Before we can ask that, however, we must ask: What is heat?

We all experience heat and take it for granted. It reaches us primarily from the sun, which is why we can feel heat in the sunlight that we don't feel in the shade. To a lesser degree, we can experience heat emerging from fire, electric light bulbs, radiators, or kettles of hot water, and even if we don't know what it is, we know what it does: It flows from one body to another. If we are cold and stand before the fire, heat passes from the fire to us. If

21. HOW DO WE MEASURE TEMPERATURE?

We can easily tell when one substance is hotter than another simply by touching and comparing them, but our sense of touch is not accurate enough to gauge exactly how *much* hotter it is. In fact, there is a familiar experiment that goes like this: Dip one hand in water that is quite hot and another in water that is quite cold and let them remain there for a while. Then dip both hands in water that is lukewarm. The lukewarm water will feel cold to the warm hand and warm to the cold hand.

In short, you are no better off judging temperature by feel than judging length by eye. You would want a meterstick for measuring lengths and a similar tool for measuring temperature. You would want some phenomenon that changed regularly with rises and falls in temperature and then marks these changes off into convenient units. Galileo was the first to try to devise such a tool. In 1603, he inverted a glass tube of heated air into a bowl of water. As the air cooled, it contracted and drew water up into the tube. When the room was warmer, the air in the tube expanded and the water level went down; when the room was colder, the air in the tube contracted and the water level went up. By measuring the water level, one could judge the temperature of the room.

Galileo's device was the first crude *thermometer* (from the Greek words meaning *to measure heat*) and also the first scientific instrument to be made of glass. It was not a very good thermometer, however, for it was open to the air, which meant that the water level in the tube was also affected by air pressure, and that confused the results. In 1654, the grand duke of Tuscany, Ferdinand II (1610–1670), invented a thermometer that was unaffected by air pressure. A liquid was sealed into a sizable bulb from which a thin tube extended, neither of which contained any air. Liquids also expand as the temperature goes up and contract as the temperature goes down. They don't expand and contract as much as air does, but even a slight expansion or contraction can cause a sizable change in the liquid's level in a thin tube.

The first liquids used for the purpose were either water or

alcohol, but neither was satisfactory. Water froze solid and would not serve for temperatures on cold winter days; alcohol boiled too easily and would not do for measuring the temperature of hot water. In about 1695, the French physicist Guillaume Amontons (1663–1705) suggested the use of mercury, which was an ideal liquid for the purpose. It remained liquid over a far greater range than either water or alcohol and seemed to expand and contract with smooth regularity as the temperature changed.

A German-Dutch physicist, Gabriel Daniel Fahrenheit (1686–1736), devised a thermometer, in 1714, in which a thread of mercury expanded upward into a thin tube containing a vacuum from a reservoir in a bulb filled with mercury. He made a mark on the thermometer at the mercury level when it was put into melting ice and another when it was put into boiling water. He divided the length between into 180 equal steps that are now called Fahrenheit degrees. The temperature of melting ice is 32°F, and that of boiling water is 212°F. There is some argument as to just why Fahrenheit chose those numbers, but he did.

The Swedish astronomer Anders Celsius (1701–1744) devised what is now called the *Celsius scale* in 1742. The freezing point of water is set at 0°C, and the boiling point at 100°C. The Celsius scale is neater than the Fahrenheit scale, and the whole world uses it now, except the United States, which clings, perversely, to the older system. It is not difficult to change from one scale to the other, however, and in this book I will give temperatures in both scales.

---

22. WHAT IS ENERGY?

Heat is merely one form of what scientists call *energy*, which is the name given to any phenomenon that has the capacity to do work. The very word *energy* is from Greek words meaning *containing work*. We must be careful, though, for scientists use the term *work* in a highly specific way that does not correspond to the way it is used in ordinary life. Work, to scientists, is what we have when we exert force over a distance against a resistance.

If we lift an object with mass straight upward for a meter against the resistance of the pull of gravity, we are doing work on the mass in the scientific sense. If we hold it motionless at a height of a meter, we are no longer doing work. We may think we are, for we grow tired, but that is only because our muscles are expending energy in keeping tense; they are doing no work on the mass. If you were to put the mass on a ledge that is a meter high, the ledge would hold the mass at that height for an indefinite period without getting tired, and it would not be doing work, either. Nor do I do work in thinking up the arrangement of words that go into the construction of this book, though it certainly tires me after a while.

Heat will do work in the scientific sense. It will expand mercury, for instance, lifting some of it against gravity. Your muscles will lift a mass. A magnet will lift an iron nail. Electricity, light, sound, and chemicals can be made to do work under the right circumstances. So can any object that is already in motion and that therefore possesses *kinetic energy* (Greek for *energy of motion*), or an object that is positioned at a height and that has a capacity to fall and do work in that fashion, as weights do in turning the handles of a clock as they fall. (An object at a height is said to have *potential energy*.)

Are all of these different kinds of energy independent, or do they have some relationship to one another? Electric current will produce a magnetic effect, and magnetism can create an electric current. An electric current can produce sound in a bell, light and heat in an incandescent bulb, and motion in a motor. Light can produce electricity, and sound can produce motion. In fact, any form of energy can be converted into any other form of energy; it is a single phenomenon that can assume a variety of forms.

But when one form of energy is converted into another form, is any of it lost in the process? Or is any of it lost when it remains in just one form? The answer to these questions seemed in earlier times to be a resounding yes. The kind of energy that was the most familiar and the most studied was the kinetic energy of motion.

A massive cannonball moving quickly enough to batter down the walls of a castle is certainly an expressive demonstration of energy. But if the cannonball is allowed to move across the ground, it does not roll on forever. It gradually slows down and finally comes to a halt, and as the motion slows, the energy content of the cannonball decreases. What happened to it? As nearly as any one could tell, the energy simply disappeared.

It took time for scientists to realize that whenever energy seemed to disappear, it was converted to heat. The rolling cannonball's energy was converted to heat as it rolled, but the heat was spread over such a long strip of ground that it was unnoticeable. If such heat is taken into consideration, is there any loss of energy in the conversion of one form of energy into another?

The first person to put this problem to a careful and persistent series of tests was the British physicist James Prescott Joule (1818–1889). During the 1840s, he conducted innumerable tests in which he converted one form of energy into another. He measured the original energy content and the energy that was produced, including heat, and concluded that no energy was either lost or gained in these processes. He described his experiments and conclusions in 1847, but he was an amateur scientist (he was a brewer by profession) and wasn't taken sufficiently seriously.

That same year, however, a German physicist, Hermann L. F. von Helmholtz (1821–1894), also announced this same conclusion. He was a professor and his analysis of the theory was so carefully done that it attracted attention. Therefore he is usually considered to be the man who first pronounced *the law of conservation of energy*, which states that energy can be neither created nor destroyed, though it can change its form. Another way of putting it is that "the total energy content of the universe is constant." Some people regard this concept as the most fundamental of all the laws of nature.

Since studies of energy are usually reduced to the study of the flow of heat, the science of the interchange of work and energy is called *thermodynamics* (from Greek words meaning *the movement of heat*). The law of conservation of energy is sometimes called *the first law of thermodynamics*.

The importance of this law (and of all similar laws) is that it sets limits on what is possible. No matter what phenomenon is observed or suggested, the question must be asked: "Where does the energy come from and where does it go?" If this question cannot be answered, then there is something wrong; either an assumption is unjustified, or an observation is mistaken, or information is incomplete.

On the other hand, the law of conservation of energy, and other similar grand generalizations, cannot be proved. All we can say is that scientists have not yet been able to find any exceptions. Exceptions may suddenly and unexpectedly arise that may force us to rethink this law, to alter, broaden, or replace it. But after a century and a half, the law of conservation of energy has held firm.

Still, even the most steadfast laws of science are subject to revision. As late as 1900, the status of nuclear energy was unappreciated, and all considerations of energy conservation were incomplete without it. Also, the notion that mass itself is a very concentrated form of energy was not yet understood, and there, too, knowledge of energy conservation was incomplete. Physicists did not feel the loss because nuclear energy and the energy of equivalence of mass happened to play no significant role in scientific investigations of the 1800s. We must therefore understand that even today there might be crucial aspects of the universe we know nothing about, which, once learned, will force us to recast our notions. Nothing in science is ever beyond improvement or modification, not even the law of conservation of energy, which is one of the traits that makes the game of science so overwhelmingly interesting.

23. IS IT POSSIBLE TO RUN OUT OF ENERGY?

The law of conservation of energy makes it clear that energy cannot be destroyed. That makes it sound as though we will always have energy available to do all the work we wish. After all, since using energy doesn't destroy it but merely, at most, changes its form, we can (we might suppose) use it in its new form, whatever it is, change its form again, and so on indefinitely.

Unfortunately, it doesn't work that way. It is the experience of scientists that every time energy is used to perform work, only part of the energy can be so used. The rest changes into heat. We can then use the heat to perform work, but only when it is unevenly distributed, when there is a region that is hot and another that is cold. When this difference is used to do work, again only part of the energy we are dealing with turns into work; the rest is lost as heat that is more evenly distributed than before. Once a quantity of heat is completely evenly distributed, no further work can be obtained from it. The result is that whenever we make use of energy to do work, we end up with energy that is less usable for further work. Energy, as a whole, cannot be destroyed, but the *free energy*—that portion of the energy that can be made to do work—continually decreases.

Another way of looking at it is to consider that all energy can do work only when it is unevenly distributed, and not just heat alone. Every time we do work, we distribute the energy a little more evenly. The measure of the even distribution of energy (which can also be defined as the *disorder* of a system—the more evenly distributed, the more disorderly) is called *entropy*. The higher the entropy, the less work we are going to get out of the energy. In this way, we might visualize the universe as slowly, but inexorably, running down.

Everything we do raises the entropy of the universe. In fact, everything that happens in the universe, even when human beings have nothing to do with it, raises the entropy. This perpetual and unavoidable increase in entropy is usually called *the second law of thermodynamics*.

The first person to get a hint of the existence of this law was a French physicist, Nicholas L. S. Carnot (1796–1832), who in 1824 published a small book on his studies of the ability of steam engines to convert the uneven distribution of heat into work. The matter was studied in detail, however, beginning in 1850, by the German physicist Rudolf J. E. Clausius (1822–1888), who first envisioned the universe running down.

But if free energy is always declining and entropy is always increasing, how is it that, although the universe has certainly been in existence for billions of years, it hasn't run down already? The answer is that the supply of free energy with which the universe started was so vast that even after billions of years not much of a dent has been made in it. The universe may be running down, but it will take many more billions of years, and there is no need for immediate concern—on that score, at least. In addition, though we know much more today about the ultimate end of the universe than Clausius did a century and a half ago, we still don't know as much as we would like to. We are not as certain now about the ultimate running down of the universe as Clausius and his followers were then.

24. WHAT IS THE INTERNAL TEMPERATURE OF THE EARTH?

Now it is possible to get back to the Earth and ask further questions, such as what is the temperature of the Earth deep below its surface?

There has always been a tendency to think of the Earth as being hot under the surface. After all, there are hot springs here and there on Earth, and there is the ferocious evidence of volcanic eruptions. It might be volcanoes that gave early human beings the notion that the interior of the Earth contains hell, a region of never-ending fire in which the souls of people you don't like are tortured forever by a vengeful and unforgiving deity.

There is no evidence for hell existing in the depths of the Earth, but there is evidence that the Earth's center is a region of great and, apparently, everlasting heat. Once human beings begin to dig deep into the bowels of the Earth for such things as gold and diamonds, it became quickly apparent that the further down people went, the higher the temperature. In the deepest mines the temperature is all but unbearable, even with air-conditioning.

Judging from the rate of rise of temperature with depth, it is reasonable to suppose that the center of the Earth might be at a temperature of 5000°C. (9000°F.)

But now that we know about the law of conservation of energy, we have the obligation to ask: Where did the energy come from that produced the heat? We will answer this question, but not till later in the book, when we consider how the Earth was formed.

25. WHY DOESN'T THE EARTH COOL OFF?

Even if we agree to leave for later the question of where the Earth's internal heat came from, it is still possible to ask why the Earth still possesses that heat. After all, if it is 4.6 billion years old, why hasn't it cooled off long ago?

By the laws of thermodynamics, heat should always flow from a region of high temperature to a region of low temperature.

It should therefore flow from Earth's hot center to its cool surface and from there out into space.

To be sure, Earth receives heat from the sun, which balances the loss of heat from the Earth's surface, but even with the sun's heat added, the temperature of the Earth's surface is, on the average, about 14°C (57°F), and what is that compared with 5,000°C (9,000°F)? Heat should still flow outward from the superhot center until the entire planet was at the same temperature as the surface. Although the rocky surface of the Earth is a good heat insulator—that is, heat flows through it only slowly, which greatly retards the process of cooling the interior—it does not reduce the flow to nothing. It certainly seems that in 4.6 billion years there has been plenty of time for the Earth to cool off, and yet it remains a very hot body. Why?

It might be that the laws of thermodynamics are wrong, but scientists don't want to have to assume that except as a last resort. First, they are bound to think that there might be a source of energy available to the Earth that they haven't taken into account, and this turned out to be true.

After radioactivity was discovered, the French chemist Pierre Curie (1859–1906), the husband of Madame Curie, realized that when radioactive atoms break down, they must release energy. In 1901, he was the first to measure this energy and found, in essence, that when an atom of radioactive material breaks down, it gives off much more energy than a molecule of gasoline does when it burns, or a molecule of TNT when it explodes. Thus, a powerful new form of energy, *nuclear energy,* whose existence scientists had never suspected, was discovered.

Radioactive materials give off this energy so slowly that in the ordinary course of events it is never noticed, but they continue to do this for incredible lengths of time. In the 4.6 billion years the Earth has existed, only half its original uranium content and only one fifth its original thorium content has broken down. In the process, the uranium and thorium present in Earth's rocky layers have produced heat that has been added to Earth's supply and kept it from cooling down. If anything, the Earth has heated up a bit, an effect that will continue for billions of additional years, though it is very slowly diminishing.

26. DOES THE SKY TURN IN ONE PIECE?

It is time to turn to the rest of the universe, for that will help us ask a few more questions about the Earth.

In ancient times, the Earth was considered to be the entire universe (in the natural sense, leaving out heaven, hell, and other supernatural realms for which there is no scientific evidence). All that existed aside from the Earth was the sky—blue by day, when the sun shone in it, and black by night, when the moon and numerous stars shone in it. (The moon is sometimes in the daytime sky, too, and can be seen dimly, despite the glare of the sun.)

The sky seemed to be and was therefore thought to be a solid vault enclosing the Earth. To primitives who thought the Earth was flat, it was a flattened semi-spherical lid that came down to the horizon everywhere at the ends of the Earth. To those who thought the Earth was a sphere, the sky was a larger sphere enclosing it, but was still thought to be a thin, solid vault. The biblical term is *firmament,* and the prefix *firm-* shows that it was thought to be solid; it is the translation of a Hebrew word that means a thin, metallic sheet.

If the sky were a solid substance, metallic or otherwise, it would have to turn in one piece, so that everything on it would have to turn in unison. But is that so?

These days, people don't look much at the sky because the large cities in which so many of us live are so garishly lit up at

night that the sights of the sky are largely drowned out. In ancient times, however, the world was truly dark at night, and when the night sky was clear, particularly when the moon was not in the sky, it was a glorious sight of twinkling stars. Sailors watched the night sky so that they could guide their ships by the stars; astrologers watched it because they thought that certain details about the sky gave hints about the future of nations and individuals; and some people watched it simply for its beauty. Those who watched the stars from places like Greece, Babylonia, and Egypt noticed that all the stars wheeled in a circle about the North Star. Stars that were close to the North Star made small circles and remained in the sky all night every night. Stars that were farther away dipped below the horizon as they circled but then emerged again later. The important point was that they all seemed to move together. They made up patterns in which imaginative people could trace animals and other configurations (the *constellations*), which remained utterly unchanged from night to night for as long as people watched. It was as though the stars were shining little spangles that were pasted onto the solid sky so that they all turned together and remained fixed in place as they did so.

For this reason they were referred to as the *fixed stars*. There are about six thousand of them visible during the course of the night to people with good vision. A few are quite bright, but most are rather dim. But why were they called fixed stars? If all the objects in the sky are fixed in place, why not take this condition for granted and simply call them stars? The trouble with that is that a few objects in the sky are not fixed in place.

One of these is the moon, which is the most conspicuous object in the night sky and must have been carefully studied even by prehistoric people. The moon rises in the east and sets in the west, as the stars do, but it lags behind them. It is impossible not to notice that it changes position in the sky relative to the stars, drifting steadily from west to east, making a complete west-to-east turn around the sky in 29½ days.

The sun does the same thing, but more slowly. Of course, you can't see the position of the sun among the stars when it is shining in the sky, for then the sky is blue and no stars are visible. But once the sun sets, the stars appear, and if you watched the situation from night to night, you would notice that the constellations all shift a little bit westward at each successive sunset. The easiest way to explain this phenomenon is to assume that the sun, like the moon, is drifting west to east against the stars, making a complete circuit of the sky in 365¼ days.

The sun and the moon are special bodies; they don't look like the other objects in the sky, but are shining discs of light rather than shining points. There are, however, five objects that *do* look like stars (although unusually bright ones) that shift position among the remaining stars. These five objects were first studied by the ancient Sumerians about 3000 B.C., and they seemed so unusual that they were given the names of gods. That habit persisted and was adopted first by the Greeks and then the Romans. We still use the Roman god names today, referring to these five starlike objects as Mercury, Venus, Mars, Jupiter, and Saturn. These five objects, plus the sun and the moon, are the wandering stars, and were called *planets,* from the Greek word meaning *to wander.* (Nowadays, we no longer call the sun and the moon planets, for reasons I will explain later.)

The seven planets have always been of absorbing interest to human beings, because their wanderings are considered by the unsophisticated to be a sort of code representing divine messages about the future (a fact exploited by unscrupulous astrologers who work out worthless messages for money). The seven-day week was invented by the Babylonians to memorialize the seven planets, and to this day, in many European languages the individual weekdays are named for the individual planets. We have Sunday, Monday, and Saturday (Saturn). The other four days are named for Norse gods. In French, however, the other four days are *mardi*

(Mars), *mercredi* (Mercury), *jeudi* (Jupiter), and *vendredi* (Venus). The ancient Hebrews adopted the Babylonian week and tried to put a religious cachet upon it in the first two books of Genesis, but the names still show their pagan origin.

Since the seven planets wander freely across the sky and can't be fixed to the solid vault of the sky, the ancient Greeks reasoned that each of the seven planets must be fixed to a sphere of its own that turned between the sky (the outermost sphere of the stars) and the Earth. Since those inner spheres weren't visible, they were assumed to be perfectly transparent and were called *crystalline spheres* from a Greek word meaning *transparent.* So the answer the ancients had to the question of whether the sky turned in one piece was yes, but with just a few exceptions, though as we shall see in due course, they were quite wrong.

27. IS THE EARTH THE CENTER OF THE UNIVERSE?

This is another question that to some there seems no point in asking. To the people of ancient and medieval times, it seemed self-evident that the Earth was the center of the universe. After all, the entire universe, to their way of thinking, was made up of the Earth and the sky. It seemed as if the sky was always above us, at about the same distance everywhere, curving with the Earth as the Earth's surface curved. It enclosed the Earth, which was at the center, so where was the difficulty?

The only uncertainty was the matter of the planets. Exactly where are they between the sky and the Earth? Since they travel at different speeds, the Greeks assumed that the faster a planet seemed to move against the background of the stars, the closer it must be to the Earth. This is the result of common experience: When we watch horses racing around a track, when they are at the far side of the track, they appear to be moving rather slowly, but when they are on the near side and gallop right past us, they seem to rush by like the wind. You can observe the same effect by watching an auto race. Likewise, an airplane traveling low in the sky appears to move along rapidly, even though the same airplane, flying much higher at the same speed, seems to pass by very slowly.

Judging by the speed of motion, then, the Greeks decided that, of the planets, the moon was closest to the Earth. Beyond it were Mercury, Venus, the sun, Mars, Jupiter, and Saturn in that order. Each one had its own crystalline sphere, seven in all, and beyond them was an eighth sphere, to which the fixed stars were attached.

It was a very pretty picture, but it did not solve the problem of the planets entirely. The ancients had to know the exact motion of the planets if they were to make astrology work. Astrologers (most of whom in ancient times were quite sincere in their beliefs) had to study those motions very carefully and, in doing so, they gave rise to the real science of the stars, which is called *astronomy*.

Even prehistoric people studied the sky closely. Thus, the stone monument known as Stonehenge in southwestern England, which came into being about 1500 B.C., might have been a device for working out the future motions of the sun and the moon.

The stars move steadily and regularly, and if the planets did, too, there would be no problem in working out their future positions (and there would be no astrology, for the code represented by their movements would be too simple to worry about). But the planets do not move in a steady and regular fashion. The moon moves a bit more slowly during one half of its course through the sky than during the other half, and the same is true, though to a lesser extent, of the sun.

The other planets are even worse. In general, they drift west to east (an effect known as *direct motion*) against the background of the fixed stars. Every once in a while, though, their motion stops, and then for a while they actually drift backward, east to west (which is an effect known as *retrograde motion*, from Latin words meaning *step backward*) before resuming direct motion. Each of the planets has its own pattern of direct and retrograde motion, and each one is brighter at some times than others.

These patterns greatly complicated methods of calculating where particular planets would be at particular times in the future. A number of Greek astronomers worked out ways of dealing with planetary motions by supposing that different planets moved on small spheres, the centers of which moved on larger spheres, some of which were slightly off-center, and so on. It was all very complicated, indeed, and the system was summarized in a book written in about 150 by a Greek astronomer, Claudius Ptolemaeus (c. 100–c. 170), usually known as Ptolemy. The mathematical structure of the universe, with the Earth at the center and various systems of spheres surrounding it, is called the *Ptolemaic universe* in Ptolemy's honor, or the *geocentric universe* (from Greek words meaning *earth-centered*). It was accepted by just about everybody for 1,700 years, and hardly anyone ever dreamed of questioning it. But it was quite wrong, just the same.

28. ONCE AGAIN, THEN, IS THE EARTH THE CENTER OF THE UNIVERSE?

There were, indeed, a few daring thinkers who questioned the general acceptance of the Earth as the center of the universe. The first person we know of by name who supposed that the Earth was not at the center of the universe, but was moving through space around some other object, which *was* the center, was the Greek philosopher Philolaus (480–? B.C.). About 450 B.C., he suggested that the Earth, along with all the planets and the sun, moved about an invisible central fire, which we couldn't see except for its reflection in the sun. This was a suggestion with no evidence or logical reasoning behind it, and no one took it seriously.

A century later, about 350 B.C., a Greek astronomer, Heracleides (388–315 B.C.), made a more sensible suggestion. He noticed that the planets Mercury and Venus never moved very far from the sun, but moved away some distance, then returned, moved away some distance in the other direction, then returned again, and repeated the process over and over. He suggested, therefore, that what was happening was that Mercury and Venus turned, or revolved, about the sun, and that the sun, carrying Mercury and Venus with it, revolved about the Earth. This made a lot of sense, but it was unacceptable to most of the Greek astronomers, who were committed to the principle that Earth was at the center of the universe and that everything revolved about it, with no exceptions.

Then about 260 B.C., another Greek astronomer, Aristarchus (c. 310–c. 230 B.C.), came up with an even more radical suggestion. This idea arose out of his attempt to determine the distance of the sun. When the moon is exactly at the half-moon stage, the moon, Earth, and sun are at the extreme points of a right triangle (I'll have more to say about this a bit later). This is exactly the kind of triangle that trigonometry deals with; if you know the exact size of the angles of the triangle, you can use trigonometry to determine how much farther away the sun is than the moon. Unfortunately, Aristarchus had no devices that would have made it possible for him to measure the angles accurately, and his estimates were way off. Even so, he decided that the sun must be twenty times as far from the Earth as the moon is, and since the sun seems to be the same size as the moon in the sky, even though it is twenty times farther away, it must also be twenty times wider.

From this information he estimated that the diameter of the sun was seven times that of the Earth, which was a serious underestimate of the situation, but it was enough to convince Aristarchus that it was ridiculous to suppose that a huge sun would circle a small Earth. He suggested, instead, that the Earth and all the other planets revolved about the sun.

Aristarchus was the first person we know of who suggested that the sun, and not the Earth, was the center of the universe (this idea is known as a *heliocentric universe* from Greek words meaning *sun-centered*), but it did him no good. Hardly any astronomers took this notion seriously.

Nevertheless, as the centuries passed, astronomers grew a little frustrated with the complex mathematics needed to understand the geocentric universe. In 1252, King Alfonso X of Castile supervised the formation of new planetary tables, which were called the *Alfonsine tables* in his honor, and he said in exasperation, "If the good Lord had asked my advice at the time of the Creation, I would have suggested a simpler system of the universe." In the 1500s, it occurred to a Polish astronomer, Nicolaus Copernicus (1473–1543), that there *was* a simpler system of the universe—the heliocentric universe suggested by Aristarchus.

Aristarchus had merely formulated the idea; he hadn't done anything with it. Copernicus, on the other hand, followed it up, and showed that a heliocentric system could account for the retrograde motions of the planets without trouble, and could also explain why they grew dimmer and brighter over time. More important still, the heliocentric system made it considerably simpler to calculate planetary tables.

Copernicus hesitated to publish his work because he knew he would get in trouble with religious leaders who were committed to the geocentric view, believing, as they did, that the Bible upheld it. His handwritten manuscript circulated among astronomers, however, and in 1543, the year of his death, it was published. (Even in a heliocentric universe, the Earth is not entirely deposed as a center; the moon, at least, circles the Earth.)

The first person to use the heliocentric universe to calculate planetary tables was a German astronomer, Erasmus Reinhold (1511–1553). These were published in 1551 under the sponsorship of Albert, Duke of Prussia, and were therefore called the *Tabulae Prutenicae* (Prussian tables). But though they were much better than the Alfonsine tables, now three centuries old, the old guard would not give up. Most astronomers refused to abandon the geocentric universe because they could not believe that the Earth was flying through space. Some maintained that even if the helio-centric universe produced better planetary tables, it was merely a cute mathematical device, which didn't mean that the Earth was *really* moving about the sun.

 The dispute continued for half a century until Galileo and his telescope settled it. In 1610, he looked at the planet Jupiter and found that the telescope expanded it into a little orb of light. That sight was the first indication that Jupiter might actually be a world. What's more, it had four lesser worlds that clearly circled it, the way the moon circles the Earth. Such subsidiary worlds were named *satellites* from a Latin term for the sycophantic followers of important people. The moon is Earth's satellite, and Galileo had discovered four satellites of Jupiter.

The importance of this discovery at the time was that it showed that four objects at least were definitely *not* circling the Earth, but were circling Jupiter, instead. This meant that Earth was certainly not center of *everything*.

Of course, it might be argued that Jupiter circled the Earth, carrying its four satellites with it, but then Galileo studied the planet Venus. By the old geocentric theory, if Venus was a dark body that shone only by reflected light, its position between the sun and the Earth was such that it should always appear as a crescent. If the heliocentric theory was correct, then Venus should show all the phases that our moon does, from crescent to full, which, as Galileo found, is exactly what Venus does.

This discovery just about established the heliocentric system. The planets, including the Earth, circled the sun, and the term *planets* was retained only for such bodies. In other words, the sun was not a planet; it was the center. The moon was not a planet, for it circled the Earth. The Earth, however, *was* a planet. Conse-quently, with the sun at the center, there were now six known planets revolving around it in the following order: Mercury, Venus, Earth (and the moon), Mars, Jupiter (and four satellites), and Saturn. All these bodies taken together came to be known as the *solar system* (from *sol,* the Latin word for *sun*).

The adherents to the old system tried to argue that everything one saw through the telescope was an optical illusion, but that only aroused laughter. In 1633, the Catholic Church resorted to force and made Galileo say (under the threat of torture) that the Earth didn't move, but that wasn't evidence. The picture of the solar system as made up of planets (including the Earth) circling the sun has been accepted by all educated people ever since Galileo's time.

Of course, the heliocentric system did raise some questions. The sun seems to follow a path across the sky at an angle to Earth's equator, thus producing the seasons. How does this work in a heliocentric system? If Earth's axis was perpendicular to the plane in which it circled the sun, the sun would seem to move across the sky right over the equator. But the Earth's axis is tilted at 23.5° to the perpendicular, a slant that remains constant as the planet revolves around the sun. This means that during one half of Earth's orbit, the northern end of the axis leans toward the sun so that the noonday sun shines north of the equator, and during the other half, the northern end leans away from the sun so that the noonday sun shines south of the equator. This view exactly explains the apparent rise and fall of the noonday sun and the cycle of the seasons.

The major divisions of time measurement can now be seen in their true astronomical form: the day is the period of the Earth's rotation on its axis; the month is the period of the moon's revolution about the Earth; and the year is the period of the Earth's revolution about the sun.

29. CAN COPERNICUS' VIEW BE IMPROVED?

Any scientific view or theory can be improved; in science the search for improvement is never-ending. Actually, Copernicus' view is not very different from that of Ptolemy. It merely changes the center of the universe from the Earth to the sun; around the sun are still the old crystalline spheres. Instead of the Earth being surrounded by seven planetary spheres with an eighth as the outermost sphere of the stars, it is the sun that is surrounded by six crystalline spheres with a seventh outermost one for the stars, while the moon has a special subsidiary sphere of its own, enclosing the Earth. The calculation of tables remained quite complicated, and while the results were obtained with greater ease and more correctness than before, many difficulties remained.

The Danish astronomer Tycho Brahe (1546–1601) spent considerable time studying the positions of the planets. He built the first important astronomical observatory of modern times and designed instruments to determine planetary positions. He did not have telescopes, however, which had not yet been invented. Nevertheless, he pinpointed the positions of planets, especially Mars, better than anyone ever had before. He believed that his calculations might make it possible to produce more accurate planetary tables, and though he died before he could produce these tables, he left his figures to his disciple, the German astronomer Johannes Kepler (1571–1630).

Kepler spent years working on the figures and found no circles that would fit the planets' positions properly. Then it struck him that if he used an ellipse instead (a kind of elongated circle), it matched the positions of Mars remarkably well. The ellipse, like the circle, has a center, but it also has two spots, called *foci,* on either side of the center along the longest diameter. These foci are located in such a way that the sum of the two distances from any point on the ellipse to the two foci represents the same length. The more elongated the ellipse, the farther the two foci are from the center. Kepler was able to show, in 1609, that each planet traveled about the sun in an ellipse, with the sun at one of the foci of the ellipse. As for the moon, it traveled about the Earth in an ellipse, with the Earth at one of the foci. This is Kepler's first law, and it meant that at one end of the planet's path (or *orbit,* from the Latin word for *circle,* even though it wasn't a circle) the planet was nearer the sun than when it was at the other end, and that the moon was nearer the Earth at one end of its orbit than at the other. Kepler's first law finally did away with the notion of crystalline spheres that had been part of astronomy for two thousand years.

Kepler also worked out a way of calculating how the speed of travel of a planet changed with its distance from the sun. The closer to the sun, the faster the planet traveled, according to a specific mathematical relationship (Kepler's second law). Then in 1619, Kepler worked out a mathematical relationship that showed how long it should take a planet to travel once around the sun if it was at a certain distance from the sun (Kepler's third law). Kepler's laws of planetary motion made it possible to work up a model of the solar system, showing exactly what kind of an ellipse each planet had and how their distances from the sun were related to each other.

Of course, if the crystalline spheres did not exist, people were bound to ask what held the planets in their orbits. Why didn't they simply wander off into space? This dilemma was answered by the English scientist Isaac Newton (1642–1727), who developed laws of motion and a theory of universal gravitation. He asserted that every object in space attracts every other object according to a simple mathematical relationship, a formula that elegantly supported Kepler's laws and explained what held planets in their orbits. The picture of the solar system as drawn by Kepler is still essentially the one we use today, and scientists are quite satisfied that no major changes will be required in the future.

30. HOW WAS THE EARTH FORMED?

Now that we have a clear picture of what the solar system is like, we can ask how the Earth was formed. We can't really consider how the Earth was formed all by itself, for, as we shall see, it had to be formed as part of a larger formation, that of the solar system as a whole. So, what happened 4.6 billion years ago that brought the Earth—and the solar system in general—into existence?

One person who took up this question without referring to the old biblical story (for which, of course, there is no scientific evidence) was the French naturalist Georges Louis de Buffon (1707–1788), who had thought the Earth might be 75,000 years old. In 1749, Buffon reasoned that the planets, including Earth, were related to the giant sun rather like chicks to a mother hen. Perhaps, then, he suggested, the Earth was born of the sun.

Buffon visualized the collision of the sun with some other large body and imagined that the impact had somehow jarred a piece of the sun loose, which cooled down and became the Earth. The suggestion was interesting, but it didn't account for the other planets, nor did it account for the formation of the sun; it assumed that the sun was simply there.

A better explanation was needed. After Kepler had produced his picture of the solar system, it was clear that it somehow represented a unit. All the planets moved in almost the same plane (so that the entire model of the solar system could be fitted into a gigantic pizza box), and they all revolved about the sun in the same direction, as did the moon about the Earth and the satellites of Jupiter about Jupiter. In addition, they all rotated on their axes in the same direction, as did the sun. Astronomers assumed that the solar system wouldn't show these similarities if it hadn't been formed all in one piece, so to speak.

The first suggestion that dealt with the formation of the solar system, rather than merely with the Earth, came as a result of the realization that there was more to the starry sky than merely stars. In 1611, in the very early days of the telescope, the German

astronomer Simon Marius (1573–1624) detected a luminous hazy patch in the constellation Andromeda. It was called the *Andromeda nebula* (*nebula* is the Latin word for *cloud*). Then, in 1694, Huygens (the inventor of the pendulum clock) detected a luminous hazy patch in the constellation Orion, the *Orion nebula*. Other nebulas were also located. (The proper Latin plural of *nebula* is *nebulae*, but in recent years the English system of adding an *s* for a plural has been adopted.)

It seemed possible that such luminous clouds were vast conglomerations of dust and gas that had not yet condensed to stars, and that all stars might have been nebulas to begin with. In 1755, the German philosopher Immanuel Kant (1724–1804) published a book that suggested just this. He thought that a nebula was slowly brought together by its own force of gravitation, swirling as it did so. The central portion became a star, and the outer portions became the planets. This proposal seemed to explain why all the planets should move in the same plane and revolve and rotate in the same direction.

In 1798, the French astronomer Pierre Simon de Laplace (1749–1827), who, possibly, did not know of Kant's earlier work, described the same idea in a book he wrote, but went into even greater detail. He pictured the nebula as slowly contracting, and as it contracted, the whirling of the nebula grew faster. This is not an idea that Laplace had to invent; the contraction is the result of gravitational pull, which we know works because it works within the solar system, and the accelerated whirling of the nebula as it contracts follows from the *law of conservation of angular momentum,* an effect every ice skater knows who goes into a spin and then spins faster when he draws his arms closer to his body.

As the nebula contracted and whirled faster and faster, its central regions bellied outward and came loose. This process, too, does not have to be made up; it is the result of the *centrifugal effect,* which is well known from observations and experiments on Earth. Laplace imagined that the portion that came loose contracted and formed a planet. The central regions continued to contract, and another planet was formed, then another, and soon all were spinning in the same direction. Finally, what was left of the central region became the sun. Because Kant and Laplace started with a contracting nebula, this conception of the formation of the solar system is called the *nebular hypothesis.* (A hypothesis is a suggestion that doesn't have the strength of evidence behind it that a theory has.)

For a century, astronomers were more or less satisfied with the nebular hypothesis, but, unfortunately, they became steadily less satisfied. The trouble came from the matter of *angular momentum.* Angular momentum measures the quantity of spin of an object, which is partially due to the object's rotation about its axis and partially due to its revolution about another object. The planet Jupiter, rotating about its axis and revolving about the sun, has thirty times the angular momentum of the sun, which is a much larger body. All the planets put together have nearly fifty times the angular momentum of the sun. If the solar system began as a single cloud with a given amount of angular momentum, how did almost all that momentum come to be concentrated in the little bits of matter that broke off to form the planets? Astronomers couldn't reach an answer, and they began to search for other explanations.

In 1900, two American scientists, Thomas Chrowder Chamberlin (1843–1928) and Forest Ray Moulton (1872–1952), returned to a version of Buffon's idea. They suggested that another star passed near the sun long ago and that the gravitational pull between them drew matter out of each. As the stars separated, the gravitational pull on this matter caused a spinning motion and set it whirling so that the planets were given a great deal of angular momentum. After the final separation, the matter condensed into small, solid objects, or *planetesimals,* and these further collided and combined into planets. The two stars, having come together singly, each departed with a family of planets. This explanation was called the *planetesimal hypothesis.*

One difference between the two theories is particularly important. If the nebular hypothesis is correct, then every star might form with planets. If the planetesimal hypothesis is correct, only stars that had experienced a near collision would have planets, and stars are so far apart and move so slowly compared with the distances between them that near collisions must be very, very rare. The difference, then, is between very, very many planetary systems for the nebular hypothesis and very, very few for the planetesimal hypothesis.

As it happens, the planetesimal hypothesis didn't survive either. In the 1920s, the British astronomer Arthur Stanley Eddington (1882–1944) showed that the interior of the sun was much hotter than anyone had expected. (There will be more to say about this later.) Material pulled out of the sun (or out of the other star) would be so hot that it could not condense into planets, but would spread out into space. The American astronomer Lyman Spitzer, Jr. (b. 1914), showed this convincingly in 1939.

In 1944, the German astronomer Carl Friedrich von Weizsäcker (b. 1912) returned to the nebular hypothesis, but improved it. He pictured the whirling nebula as developing subwhirls that formed first planetesimals and then planets. Astronomers could take electromagnetic effects in the nebula (something not known in Laplace's time) into consideration to account for the way in which angular momentum was transferred from the sun to the planets.

The formation of planets from planetesimals, by the way, explains the internal heat of the Earth. The planetesimals are moving quickly and have enormous kinetic energy, but as they smash together, they come to a halt and the kinetic energy turns

to heat. By the time a planet is formed, the heat produced by all the halting of motion is enormous, which is why the Earth's core might be as hot as 5,000°C (9,000°F). Naturally, the larger a planet, the more kinetic energy has been turned into heat in the process of its formation and the hotter it must be at its core. The smaller a planet, the less kinetic energy was involved in its planetesimals to begin with and the cooler it must be at the center. Thus the moon is undoubtedly much cooler than 5,000°C at its center, since it is smaller than the Earth, and Jupiter, which is larger than the Earth and, indeed, is the largest planet, must be considerably hotter. Some estimates place the central temperature of Jupiter at 50,000°C (90,000°F). For now, then, the new version of the nebular hypothesis seems to be quite satisfactory.

31. IS THE EARTH A MAGNET?

Since we've mentioned electromagnetic phenomena in connection with the formation of the solar system, it would seem that the bodies of that system have magnetic properties. Could it be, then, that the Earth is a magnet? Indeed, scholars have been puzzling over the matter for centuries.

The first person we know of to describe the property of some substances to attract iron was Thales (c. 636–c. 546) in about 550 B.C. He described it in connection with a piece of rock found near the city of Magnesia in Asia Minor, from which the word *magnet* was derived. Magnets remained merely a curiosity until the Chinese discovered that magnetized needles invariably oriented themselves, if given a chance, in the north-south direction. In 1180, the English scholar Alexander Neckam (1157–1217) mentioned such a magnetic compass. It was eventually used to guide ships over the ocean and led to the great European Age of Exploration beginning about 1420.

In 1269, a French scholar, Petrus Peregrinus (c. 1240–?), was the first to study magnets systematically. He found, among other things, that every magnet has two poles with opposite magnetic properties—usually referred to as the *north magnetic pole* and a *south magnetic pole*. The north pole of one magnet attracts the south pole of another, though two north poles or two south poles repel each other. But why would a magnet's north pole point to the north? Was the Earth itself a huge magnet? This possibility was investigated by an English scientist, William Gilbert (1544–1603), who shaped a lodestone, the magnetic substance Thales had first studied, into a sphere. In 1600, he published a book in which he described the way a magnetic compass acted in the neighborhood of that magnetized sphere and showed that it acted just as it does in connection with the Earth. It then seemed that the Earth must be a magnet.

But why? One possibility was that there might be a huge piece of magnetized substance at the Earth's center, pointing north and south, and once people began to speculate that the Earth might have an iron core, that seemed to be the answer. In 1895, however, Pierre Curie discovered that iron lost its magnetism at temperatures over 760°C (1,400°F), and since it is quite likely that Earth's core temperature is well over that figure, it is obvious that the core is not a magnet in the ordinary sense.

However, hot liquid iron will still conduct an electric current, and if the iron swirls, the turning current would create a magnetic field. Earth could then be, not an ordinary magnet, but an *electromagnet*. In 1939, the German-American geophysicist Walter Maurice Elsässer (b. 1904) suggested that Earth's rotation could cause swirls in the core that would produce a magnetic field.

Although this is a widely accepted thought now, problems remain. The Earth's north and south magnetic poles are not located at the geographic poles but about 1,600 kilometers (about 1,000 miles) away in each case. If a line is drawn from the Earth's north magnetic pole to its south magnetic pole through the body of the planet, it does *not* pass through the center of the Earth. What's more, the magnetic poles drift slowly, and the magnetic field grows more or less intense with time. There are times, in fact, when the field drops to zero, then reverses and grows in intensity again. There is still much we don't know about the detailed mechanics of Earth's magnetic field.

32. IS THE EARTH A PERFECT SPHERE?

As we learn more and more about the Earth, it becomes possible to ask more detailed questions. For instance, scientists have known for nearly 2,500 years that the Earth is a sphere, but is it a perfect sphere?

Why shouldn't it be? If the Earth is a sphere because the pull of gravity draws its substance as closely as possible to the center, then it should indeed be a perfect sphere. Besides, the sun makes a perfect circle in the sky at all times, and so does the moon, which means that those bodies are perfect spheres.

The first evidence to upset that notion was the sight of Jupiter and Saturn through a telescope in the early 1600s. Both seemed to be elliptical rather than circular in appearance, and they maintained that shape as they turned. What's more, the longest diameter of the planetary ellipse in those two cases seemed to be that of the equators of those planets, which means they were both spheres that bulged at the equator and were flattened at the poles. Bodies of which this is true are called *oblate spheroids*.

Why should Jupiter and Saturn be oblate spheroids?

The answer did not come until Newton worked out his laws of motion and tackled the problem in 1687. Every particle of a planet is forced to turn as the planet rotates on its axis, although the natural tendency is for an object, once it is moving, to continue onward in a straight line. There is a compromise, then; the planet turns, but the surface bulges out slightly as it does so, as though it felt the tendency not to turn but to move straight ahead. This is a *centrifugal effect* (from Latin words meaning *to fly from the center*), a phenomenon that has been studied in detail here on Earth. The faster an object rotates, the farther it bulges outward.

As a planet turns, the portions of its surface near the pole make very small circles in the course of the turn. Such points don't move fast and therefore don't bulge out much. As one moves farther from the pole, the surface makes a bigger circle, but must make the turn in the same time. The points on the surface must therefore move more quickly and bulge outward more, an effect which reaches its maximum at the equator. Thus, there is a central bulge on a turning planet, one that is deepest at the equator itself.

The size of the equatorial bulge of a world depends on how fast the surface moves and how strongly the world's gravitational pull resists the formation of a bulge. The moon, Venus, and Mercury all spin so slowly that there is no equatorial bulge worth mentioning in any of these cases. On the other hand, the sun spins quite rapidly, and a point on its equator travels at the speed of 13,600 kilometers (8,500 miles) per hour, but its gravitational pull is so intense that there is no equatorial bulge worth mentioning, either.

Jupiter and Saturn are very large compared with the Earth, and yet they rotate on their axes more rapidly. Jupiter rotates in a little less than ten hours, and Saturn, though somewhat smaller, in a little over ten hours. A point on Jupiter's equator travels 45,765 kilometers (28,400 miles) per hour, while a point on Saturn's equator travels 36,850 kilometers (22,900 miles) per hour. These are faster motions than those of points on the sun's equator, and both Jupiter and Saturn have much smaller gravitational pulls than the sun has, not enough to resist the centrifugal effect. Therefore both planets have pronounced equatorial bulges. Saturn's surface moves less rapidly than Jupiter's, but Saturn also has a smaller gravitational pull, and thus has a greater equatorial bulge.

But if all this is true of Jupiter and Saturn, might it not also be true of the Earth? The Earth spins on its axis more quickly than the moon, Mercury, or Venus. A point on Earth's equator travels at a speed of 1,670 kilometers (1,040 miles) per hour, which is much less than the equatorial speeds of the sun, Jupiter, and Saturn, but Earth's gravitational pull is much less, too. It seemed to Newton that the Earth's equatorial bulge ought to be large enough to measure.

The way to test this theory would be to go to various parts of the Earth and make careful measurements of distances and angles so that we could tell how much the surface of the Earth curved. If the Earth was a perfect sphere, it would curve equally everywhere; if it was an oblate spheroid, it would curve more near the equator than near the poles. In 1736, a French expedition under Pierre Louis de Maupertuis (1698–1759) went to Lapland, near the North Pole, to measure the curvature there. At the same time, another French expedition, under Charles de La Condamine (1701–1774), went to Peru, near the equator, also to measure the Earth's curvature.

Newton proved to be right. Earth had an equatorial bulge, though not much of one. The Earth's equatorial diameter is 12,756 kilometers (7,926 miles), while its polar diameter is 12,713 kilometers (7,900 miles). The difference is 43 kilometers (26 miles). In other words, the Earth is *almost* a perfect sphere, but not quite.

In 1959, a satellite called Vanguard I was placed in orbit about the Earth by the United States. From the precise manner in which it circled the Earth, it could be calculated that the equatorial bulge was 7.6 meters (25 feet) higher south of the equator than north of the equator. This was announced as proof that the Earth was pear-shaped; that is, broader in the southern half than in the northern. It was an unfortunate remark, for the difference in the bulge is measurable only under the most delicate conditions. In fact, the entire bulge is so small as not to be visible at all to the unaided eye, and to anyone looking at Earth from space, it would appear to be a perfect sphere. Calling the Earth "pear-shaped" made some people think that as seen from space, the Earth was shaped like a Bartlett pear, which is an utterly false picture of the situation. Fortunately, the use of the expression quickly died.

33. WHY DOES THE MOON CHANGE ITS SHAPE?

It is time, now, to turn our attention to other parts of the universe. The ancient Greeks felt, quite correctly, that the moon was, of all objects in the sky, closest to Earth, so it seems fitting to consider it next.

The moon is the only permanent object in the heavens that visibly changes its shape from night to night. The sun is always a blindingly bright circle of light. The other planets and stars are always points of light. Some comets do have peculiar and changeable shapes, but they are not permanently visible in the night sky. (I'll talk about them later in the book.)

The moon, however, goes through a steady, progressive, and repetitive series of changes. At

some particular night the moon appears low in the western sky just after sunset as a very thin crescent. From night to night it drifts eastward and the crescent grows thicker. After about a week it is a semicircle of light and continues to get larger, until after another week it is a full circle of light. Then it begins to shrink. It is back to a semicircle a week later (but it's the other half of the circle that is now lit up) and, finally, after still another week, it is a very thin crescent in the eastern sky just before dawn. After that, it disappears for a couple of nights and then the whole cycle starts over again.

The natural thought might be that the moon is like a living being. It is born, grows, reaches maximum size, then fades and dies, going through all these stages in one month. Even today, we still talk of the thin crescent in the western sky as a *new moon* and the thin crescent in the eastern sky a month later as the *old moon*. Halfway between is the *full moon*. As I explained earlier, this cycle of phases of the moon established the *month* as a unit of time, and the first calendars were based on this. But why did the phases take place? Was a brand-new moon really born each month? The Greek philosopher Thales didn't think so, and before him the Babylonian astronomers probably didn't think so, either.

The reason for this skepticism came from a consideration of the position of the moon relative to the sun as the month wore on. To begin with, it seemed natural to suppose that the rules that governed things on Earth were different from the rules that governed things in the sky. On Earth, everything fell downward; in the sky, everything traveled in circles. On Earth, everything changed and decayed; in the sky, everything seemed permanent and unchanging. The substance of the Earth was dark; in the sky, every object glowed unceasingly. If the material that made up the moon glowed unceasingly, as the sun, the planets, and the stars did, then the moon would always be an unchanging circle of light. Since the moon was not an unchanging circle of light, then either it grew and decayed in the course of the month or it did not glow unceasingly. If the moon was indeed as dark as the Earth and glowed only by reflecting sunlight, then different parts of it would reflect sunlight, depending on where the moon and sun were in relation to each other.

For instance, if the moon is just about between the Earth and the sun, then the sun would be shining on the side of the moon away from us and we would not see anything of the moon. However, the moon moves twelve times as fast as the sun does west to east, so the next night it is a little to the east of the sun and we can see just a tiny sliver of the lighted side on its western edge. It appears to be a thin crescent. As the moon continues to move eastward, we see more and more of its lighted portion, and the crescent thickens.

When it is a quarter of the way around the sky, compared with the sun, its western side is lit up and we see a semicircle of light, or a *half-moon*. This continues to grow until the moon is on the side of the sky opposite to the sun. Then the sun shines over Earth's shoulder, so to speak, and lights up the entire side of the moon that faces us, so that we see the full moon.

The moon then starts to catch up with the sun again, and the lighted portion that we can see shrinks. After a week only the eastern half is lit up, and that shrinks into a crescent. The moon then passes the sun, and the entire cycle begins again. Any person who considers this situation carefully is bound to come to the conclusion that the moon, like the Earth, is a dark body that shines only by the reflection of sunlight.

34. DOES THE EARTH SHINE?

If the moon is a dark body that shines, as it does, by reflecting sunlight, then might not the Earth, which is also a dark body, shine by reflecting sunlight, too? It seems reasonable to suppose this, but people were reluctant to accept that thought, clinging to the notion that things on Earth were fundamentally different from things in the sky. How could the Earth shine as though it were a heavenly object, when it wasn't?

Of course, the best way to find out if Earth glows like the moon would be to go far out into space and look back at the Earth from a great distance. But that was not possible until the 1960s, before which time the question had to be settled right on Earth's surface.

Oddly enough, it was. Sometimes when the moon is a thin crescent, you can see the dim reddish structure of the rest of the moon filling out the circle. It *is* the rest of the moon, and not some other body, because the moon has certain visible markings on it, and the dim reddish structure has those same markings. People still call this effect "the old moon in the new moon's arms," and for a long time no one had any good explanation of it. In about 100 B.C., the Greek philosopher Posidonius (c. 135–c. 50 B.C.) thought that the moon was partly transparent so that a little sunlight leaked through it. In about 1550, the German mathematician Erasmus Reinhold suggested that the moon wasn't entirely dark but glowed softly even when the sun didn't shine on it.

But suppose the Earth did reflect sunlight the way the moon does.

When the moon is in a thin crescent stage, it is almost exactly between us and the sun so that we see only a little bit of its lighted surface along one edge. If you were on the moon at this time, however, you would see the sun shining over the moon's shoulder, so to speak, and lighting up the entire face of the Earth that happens to be pointed toward the moon. In short, when you see a new moon from Earth, you see a *full Earth* from the moon. (In fact, if the Earth shone by reflected light as the moon does, then the Earth phase as seen from the moon would be exactly the opposite of the moon's phase as seen from Earth at that time.)

When the moon is new, the side of the moon facing toward us is getting no sunlight, but there is a full Earth in the moon's sky. The Earth is larger than the moon, and because of its cloudy atmosphere, it reflects more of the sunlight that falls on it than the moon does. On the whole, the full Earth as seen from the moon is about seventy times as bright as the full moon as seen from Earth.

The unlit side of the moon is therefore receiving the light of the full Earth. *Earthlight* is far feebler than sunlight, but it is enough to light up the dark side of the moon measurably, and it allows us to see the moon's dark side very faintly at the time of the new moon. Galileo was the first to advance this explanation of "the old moon in the new moon's arms," and it made so much sense that few have doubted it since.

35. WHY ARE THERE ECLIPSES OF THE SUN AND THE MOON?

Every once in a long while, something dark advances across the face of the sun. The sun begins to shrink and shrink and sometimes shrivels down to a thin crescent and disappears altogether. Where the sun had existed only moments before there is now only a dark circle in the sky, surrounded by a dim, pearly glow. The land grows dark, a cold wind springs up, birds start going to sleep, and human beings are sometimes frightened out of their wits. What has happened? The natural supposition is that some cosmic wolf or dragon has swallowed the sun, that it will never shine again, that a permanent cold and dark will settle down on Earth and that every kind of life, including human beings, will die. That never happens; after a few minutes, the sun begins to appear again on the side that had first disappeared. It gets larger and larger, and after a little while it is shining in full splendor.

What really happened?

The crucial point was probably first noted by the Babylonian astronomers: When there is an eclipse of the sun, the sky turns dark and the stars appear, but the moon never appears. That is because the eclipse of the sun always seems to take place at the time of the new moon, when the moon is passing the sun from west to east. There is the answer. The moon moves in front of the sun and covers it up, so that it can't be seen. Then it moves away and the sun appears again.

In that case, why isn't there an eclipse of the sun at every new moon? Why isn't there an eclipse every month? That is because the sun and the moon don't follow exactly the same path across the sky. The two paths are at a small angle to each other, and usually as the moon passes the sun, it travels a little above or a little below it. It is only when the moon passes the sun at a time that both of them are in the part of their paths that cross each other that the moon really gets in front of the sun. There are two such crossing points, or *nodes,* at opposite sides of the sky, and sometimes such a crossing takes place, causing an eclipse of the sun.

When the moon passes in front of the sun, the shadow of the moon falls on Earth's surface. (The moon casts a shadow, just as everything else does that is solid, bathed in light, and without light of its own.) The shadow that reaches Earth has narrowed to the point that only a small part of the Earth's surface is covered. It might be only 160 kilometers (100 miles) across, or sometimes much less, which means that although you might see the entire sun disappear, people a few miles away would see the moon cover only a part of the sun's face (a partial eclipse), and still farther away they would see no eclipse at all. The shadow travels across the Earth's surface as the moon moves, but altogether it covers only a small fraction of Earth's surface, and in any one place it lasts only seven minutes at most.

The sun and moon appear to be slightly different sizes in different parts of the sky. Since the moon is, on the average, a tiny bit smaller in appearance than the sun, it sometimes doesn't cover the entire sun when it moves directly in front of it. A thin ring of blazing sunlight can be seen on every side of the dark moon. This effect is known as an *annular eclipse.*

Once astronomers work out the way the sun and moon move along their paths, they can predict when an eclipse will happen. This was an important task of ancient astronomers, for people had to be prepared for an eclipse, since it was usually considered to be a message from the gods. Predicting them accurately was good business, for it made astronomers seem to be skilled interpreters of the will of those gods.

The Babylonians learned how to calculate eclipses in advance, and in Greece, Thales learned the trick from them. He is said to have predicted that there would be an eclipse of the sun in Asia Minor in 585 B.C., and there was. In fact, the armies of two nations in that region, Media and Lydia, were getting ready to fight a battle when the eclipse took place. The two armies were so frightened at this bad omen that they hastily called for peace and went home without fighting. Modern astronomers have calculated backward to get the exact day of the eclipse, which was May 28, 585 B.C. Thus, the called-off battle is the first human event in history that is known to the exact day.

Sometimes the moon is eclipsed. This only happens at the time of the full moon, when the sun is at one side of the Earth and the moon at the opposite side. If we understand the eclipse of the sun, there's no trouble with the eclipse of the moon; it happens because the moon passes through the Earth's shadow.

The Earth is considerably larger than the moon, so the Earth's shadow is considerably larger. In fact, the Earth's shadow can cover the entire moon (which is smaller than the Earth), causing a lunar eclipse that can be seen by everyone on the side of the Earth facing the moon at the time. Lunar eclipses last longer than solar eclipses.

Again, a lunar eclipse doesn't happen at every full moon

because the moon and sun follow slightly different paths. Usually, at the time of the full moon, the Earth's shadow passes over the moon or under it. Only when the sun is at one node and the moon is at the other is there an actual eclipse. Eclipses of the moon can be predicted, too, and, in fact, some people think the ancient stones of Stonehenge were arranged in such a way as to predict when these eclipses would take place.

36. DOES THE MOON TURN?

In connection with "the old moon in the new moon's arms" I mentioned the dim markings on the face of the moon, markings most clearly seen at the time of the full moon. These markings puzzled people, and imagination being what it is, some people saw them as a human being, the well-known "man in the moon," while others saw rabbits or crabs or other shapes.

To ancient scholars who thought that the heavenly bodies were changeless and perfect, the markings on the moon were a puzzle. There should be none at all; it should be a perfect field of light, as the sun was. One ingenious explanation was that the moon, being the closest of all heavenly bodies to the Earth, managed to absorb some of Earth's stains and imperfections.

Whatever the markings were, however, they always remained visible and never changed position. This seemed to mean that the moon always turned the same face to the Earth and was therefore not spinning on its axis. This isn't true, of course. To be sure, the moon always points the same face to the Earth, but it is spinning just the same.

But suppose it wasn't spinning. Suppose it revolved around the Earth in such a way that one of its faces always pointed toward a certain distant star. This means that at one point of the moon's orbit that star would be in the same direction as the Earth, but of course, far beyond it, and the moon would be facing them both.

If the moon continued moving about the Earth till it was on the other side of the planet, the face would still be pointing toward the star and would face away from us. The moon would have its back to us, so to speak, and we would be able to see its other side. In short, if the moon was really not spinning, then as it turned about the Earth, we would be able to see every part of it, little by little. It has to spin in order to show only one face to Earth.

What's more, it has to spin in a certain way: It has to make one turn in exactly the time it takes to go around the Earth. As the moon turns in this way, we see only one side of it, but it shows different sides to the sun. If you were on the moon, you would see the Earth stay pretty much in one spot in the sky (if you were on the side that always faced the Earth). However, you would see the sun move across the sky, making one complete turn in 29½ days. The moon would have a daytime that was just over two weeks long and a nighttime that was just over two weeks long. The first person to point this out clearly to the general public was Kepler, in his science fiction story *Somnium,* which was published after his death in 1634.

37. HOW FAR AWAY IS THE MOON?

The ancient Greeks decided that the moon was the heavenly body nearest to us, but exactly how close is it?

There are two things about the moon that were unknown in ancient times: its size and its distance. The two are interconnected. If the moon's size was known, it would be easy to calculate by trigonometry how distant it had to be in order to appear as large as it does in the sky. On the other hand, if the moon's distance was known, trigonometry would tell us how large it would have to be to appear the size it does. If neither quantity is known, one is stuck.

What does one do? One might go by appearances at first.

How large does the moon seem? Many people, if asked to estimate the moon's size, might say that it appears to be 0.3 meters (1 foot) across, which can't be true, of course. If the moon was really that size, it would have to be only 17 meters (56 feet) above the ground and wouldn't make it past a tall building, let alone a mountain. If the moon is going to clear the Earth's mountains, it has to be at least 9 kilometers (5½ miles) above the ground, and in that case, it must be at least 90 meters (295 feet) across.

The chances are that it is farther from the Earth than that and therefore even larger. About 460 B.C., a Greek philosopher, Anaxagoras (c. 500–428 B.C.), suggested that the sun might be a blazing rock about 100 miles across (and in that case, the moon might be quite large as well). This notion aroused such enmity in Athens, and such charges of impiety and atheism, that Anaxagoras had to leave the city hurriedly in order to save his life.

Well, then, what's to be done? There's no point in guessing. Is there any way of working out the distance of something you can't reach? Actually, there is. Hold your finger in front of your eyes and close your left eye only. You will see your finger with your right eye, and it will appear to be against the wall in front of you. Don't move your finger, but open your left eye and close your right eye. Now you see your finger with your left eye, and it seems to have shifted position against the wall. You view your finger at different angles, you see, with your left eye and right eye.

This shift of an object seen from two different positions increases as the object moves closer to you and decreases as the object moves away from you; it also increases when the object is viewed from two positions that are wide apart and decreases when it is viewed from two positions that are close together. The shift is called *parallax*. If you view a distant object from two different positions and know how far apart the positions are, and if you can measure the size of the parallax, then, using trigonometry, you can calculate the distance of the object, even if you can't reach it. Surveyors might use parallax, for instance, to tell the distance of an object on the other side of a river.

Can we apply the method of parallax to the moon? Of course,

everything shifts and shows a parallax when you change position, but very distant things show so little parallax that you might as well say they don't shift at all. Thus, if the moon is viewed from two positions a few hundred miles apart, it might shift its position slightly compared with distant stars.

This means that an astronomer could measure just how far from a particular star the moon might be at a particular time on a particular night. (The distance is measured in angles. A line making a big circle around the sky can be divided into 360 equal *degrees of arc*. Each degree of arc can be divided into 60 equal *minutes of arc,* and each minute of arc into 60 equal *seconds of arc.*) Far away, another astronomer measures the distance between the moon and the same star at the same time of the same night. The two distances are compared, and if there is a difference, then that is the parallax, and the distance of the moon can be determined.

This was first done in about 150 B.C. by the Greek astronomer Hipparchus (c. 190–c. 120 B.C.), who found that the moon's distance was equal to thirty times the diameter of the Earth. This would make the moon about 385,000 kilometers (239,000 miles) away from us, a figure that is just about correct.

This must have been a startling figure, and I doubt that anyone who heard of Hipparchus' measurement at the time could bring himself to believe it. After all, if the moon is 239,000 miles away, it must be nearly 3,500 kilometers (2,200 miles) across. That's a little over a quarter of the diameter of the Earth, so the moon would have to be accepted not just as a silver plate in the sky but as a world.

The moon's distance was as far as the ancient Greeks could go; the parallaxes of all other heavenly objects were too small to measure. Still, the moon's distance was enough to give humanity its first notion that the universe was very large and contained worlds other than the Earth.

If there was any doubt about this, it ended in 1609, when

Galileo turned his telescope on the moon. He saw mountain ranges, plains, and what looked like volcanic craters. It was these features that accounted for the markings on the moon that were visible from the Earth even without a telescope. Thus the moon was definitely a world.

38. WHAT IS THE MASS OF THE MOON?

Even if an ancient astronomer were willing to accept Hipparchus' view of the moon and to think of it as a huge world, he might argue that heavenly bodies were composed of pure light and had no substance. Their size might be no more important than the size of a cloud or a shadow, for instance.

It would be important, then, to determine the *mass* of the moon—how much matter it contains, so to speak. But how can you do that? You can't weigh it or exert force on it to change its motion, any more than you could in the case of the Earth. Nor could you go to the moon (not till 1969) to measure the pull of gravity on its surface and determine its mass in that way.

What you could do, though, is to measure the pull of the moon's gravity (assuming it has any) on the Earth itself. To work out that answer, let's think about a seesaw, something we can all visualize. Imagine a long, flat board pivoted over an axis, with a youngster sitting on either end. One youngster is down, with his feet touching the ground. He kicks, so that his end of the board flies up and the other goes down. When the other youngster reaches the ground, he kicks and the motion is reversed. This can be kept up for as long as the youngsters wish.

But suppose one youngster is much heavier than the other. The heavy youngster can kick at his end of the seesaw, which would then go up a bit and come down again, for the weight of the light youngster would not be enough to bring his end down and keep the heavy youngster in the air. The seesaw wouldn't work in that case.

The trick would then be to balance the seesaw at a point nearer the heavy youngster. The nearer the pivot is to the heavy youngster, the more difficult it is for him to keep his end down and the easier it is for the light youngster, far from the pivot, to do so. Finally, a place is found for the pivot that lets each youngster have an equal ability to bring his side down, and the seesaw is back in balance.

If you weigh the youngsters and measure the distance of each from the pivot when the seesaw is in balance, you will see that if the heavy youngster is twice as heavy as the other, the other needs twice the length of seesaw that the heavy one does. In fact, if you know the weight of one youngster but not the other, then if you measure the lengths of the seesaw on each side of the pivot when it is in balance, you can calculate the weight of the other youngster without actually weighing him. This is the *principle of the lever*, which was first worked out in full mathematical detail in about 250 B.C. by the Greek mathematician Archimedes (287–212 B.C.).

The situation of the Earth and the moon is something like that of the two youngsters on the seesaw. The Earth's gravity pulls at the moon, so that the moon revolves around it; the moon's gravity, however, pulls on the Earth, so that there is also a tendency for the Earth to revolve about the moon.

If the Earth and the moon were exactly equal in mass, then the two tendencies would be equal, and the Earth and the moon would each revolve about a point halfway between the center of the Earth and the center of the moon, with each body on the opposite side of the orbit from the other.

But if the Earth is more massive than the moon, the point about which both pivot, the *center of gravity*, has to be closer to the center of the Earth, just as the pivot of the seesaw has to be closer to the heavy youngster. The Earth is considerably more massive than the moon, so that the center of gravity is quite close to the center of Earth, so close that we can simply consider the moon to be revolving about the Earth and the Earth to be standing still.

Nevertheless, the Earth is *not* standing still. It is making a small circle about the center of gravity each month, and the center of the Earth is always on the side of the center of gravity that is away from the moon. You can tell the size of the small circle the Earth is making each month by studying the motion of the stars in the course of the month. As the Earth makes a small circle each month, the stars seem to make a small circle in the opposite direction.

The center of gravity of the Earth/moon system is 81.3 times as close to the center of the Earth as to the center of the moon. The center of the Earth/moon system is about 4,700 kilometers (2,900 miles) from the center of the Earth. This is 1,600 kilometers (1,000 miles) below the surface of the Earth, so you see that it does look as though the moon does all the revolving.

This also means that the moon has a mass that is $\frac{1}{81.3}$ (or 1.2 percent) that of the Earth. That may not sound like much, but it still means that the moon's mass is 740 thousand billion billion kilograms.

Furthermore, because the moon has a smaller mass, it also has a weaker gravitational pull. You might think that if we were standing on the moon, we would have only $\frac{1}{81.3}$ as much weight as we have on Earth, but we must remember that because the moon is a smaller body, we would be closer to the moon's center than we are to Earth's center on Earth's surface. That raises the moon's surface pull, so that if we were standing on the moon, we would have one sixth the weight we have on Earth.

Once we have the moon's mass and know its size, we can calculate its density, which turns out to be 3.34 grams per cubic centimeter, only three fifths that of Earth. From this we can deduce at once that the moon doesn't have an iron core as Earth does, but must be rock all the way through. Furthermore, since the moon is smaller than Earth, its central temperature is lower than that at the center of the Earth, and since rock doesn't melt as easily as iron does, we are safe in assuming that the moon does not have a liquid center of any kind.

Without a liquid center there is nothing at the moon's core that could swirl, and even if there were, the moon rotates far too slowly to set up those swirls, so we can conclude that the moon has no magnetic field. When probes were sent to the moon to measure its magnetic properties, it was discovered that this was correct; the Moon, unlike the Earth, is not a magnet.

As for other planets, Mars rotates fairly rapidly, but it, too, lacks an iron core. Mercury and Venus do have iron cores, but they rotate very slowly. The result is that these worlds are not magnets, either. (Mercury shows a tiny bit of magnetism, which is puzzling.)

39. WHAT ARE THE TIDES?

At any given time, the side of the Earth that faces the moon is about 7 percent closer to the moon than the side that faces away from the moon. That means that the side facing the moon gets a somewhat stronger gravitational pull from the moon than the other side. The Earth is therefore stretched slightly on a line connecting the center of the Earth with the center of the moon, and there is a bulge on either side.

The solid surface of the Earth doesn't give much, but the water of the ocean is held together less tightly and it bulges considerably more than the land does. There are thus two bulges of seawater, one facing toward the moon and one facing away from the moon. As the Earth turns, each portion of its land surface moves into the bulge of seawater and then pulls away from it.

As viewed from the land, it seems that the ocean rises higher, increasing to a maximum *high tide* and then receding to a minimum *low tide*, which it does twice a day. Actually, since the moon moves in its orbit between one high tide and the next, a given spot on the Earth's land surface is likely to experience a high tide every 12½ hours.

If this was all there was to it, human beings would have connected the tide and the moon from prehistoric times. However, there are complications. The sun also produces tides, though only about a third as high as those produced by the moon, and when the sun and moon are in the same straight line, at full moon and at new moon, the tides creep higher and recede lower than usual. When the sun and moon pull at right angles, as when there is a half-moon, the tides are less extreme than usual. Then, too, in judgments about when a tide might come and how high it will get, a lot depends on the shape of the shoreline.

The first civilizations of the West lived along the shores of the Mediterranean Sea, which is nearly landlocked. At high tide, water would flow into it from the Atlantic Ocean through the narrow Strait of Gibraltar, but long before the process was complete, the low tide had come and water began moving out. Long before that process was complete, high tide came again, so that, in the end, the change in the Mediterranean's water level was very slight.

About 300 B.C., the Greek explorer Pytheas (c. 330–? B.C.) sailed out of the Mediterranean for the first time. He cruised through the Atlantic to the British Isles and Scandinavia and, in the process, came across the phenomenon of the tides. He reported this and even suggested that it had something to do with the moon. This received little attention, however. When Julius Caesar led a raid into Britain, he beached his ships just a little way up the shore and nearly lost them to an unexpected high tide. Being Caesar, he quickly corrected his error.

The connection with the moon was difficult to accept in the absence of an understanding of gravity. Galileo, for instance, who was so unerring a thinker in most ways, laughed off any suggestion that the moon could have an influence on the Earth, and thought the tides were just the sloshing of the ocean as the Earth rotated. It was only when Newton worked out the theory of universal gravitation in 1687 that tides could be completely understood.

40. HOW DO THE TIDES AFFECT THE EARTH?

The tides are extremely important to shipping. When high tide comes in, the water in a harbor is deeper and a heavily laden ship is less likely to scrape bottom or have trouble with shoals, reefs, and rocks. Ships, therefore, tend to sail with the tide. If, for any reason, the ship can't sail at that time, it has to wait for the next high tide, regardless of how much in a hurry it is. Hence the saying "Time and tide wait for no man."

But there is a more important, if less immediate, effect of the tides. As the Earth turns through the ocean bulges on either side, there are places where the water is shallow enough to cause considerable friction between the water and land as they move relative to each other. The water scrapes the ocean bottom over large areas as it comes in to high tide and recedes to low tide.

This friction acts exactly as the friction of the brake linings of a car act. Some of the turning motion of the Earth is used up by the scraping action of the tides, and the planet experiences a braking effect. The turning of the Earth is so forceful, however, that the braking effect is minuscule. In fact, as a result of the tides, the day lengthens by only one second in 62,500 years.

Nevertheless, although the total shortening is very slow, it accumulates. Even an additional few ten thousandths of a second each year means that the shadow caused by a total eclipse of the sun, which would always fall in the same place if the day was exactly the same length to the split second at all times, falls a hundred kilometers beyond the shadow of the previous eclipse. It is from the misplacement of the early eclipses that we can work out the slow lengthening of the day.

As it happens, though, the turning of the Earth can't be braked without an effect elsewhere. The Earth has angular momentum because of its spin, which cannot be totally destroyed. If the Earth's spin grows less, then the moon's spin must grow more, so that as the day lengthens, the moon recedes slightly from the Earth and makes a larger sweep around it.

Naturally, the Earth has a tidal effect on the moon, too. Since the Earth is 81.3 times as massive as the moon, it produces a considerably larger tidal effect (even though the smaller size of the moon diminishes the effect a bit). The moon has less angular momentum than the Earth, and its spinning is more easily braked when the tidal effect causes layers of rock on its surface to strain against those below. As a result, the moon's rotation has slowed to the point that it turns only once in the course of its revolution around the Earth. That means it faces only one side to the Earth, so that the tidal bulge on its Earth-facing side and the one on the opposite side are frozen in place, and it undergoes no further slowing because of the Earth's tidal effect. Thus, it is not a coincidence that the moon's rotation on its axis and its revolution about the Earth take the same amount of time; it is a consequence of the tidal effect.

41. IS THERE LIFE ON THE MOON?

In the first place, in considering this question, we must ask what we mean by life. On Earth, all life-forms, however different they may seem in appearance, are built out of the same chemicals and have the same fundamental requirements for existence. They make up "our kind of life," or "life as we know it."

It may be that there are other kinds of life, radically different, with a different chemical background, with a different set of fundamental requirements, and with a nature so different in every way from life on Earth that we might not even recognize it as life if we were to encounter it.

We know nothing about such other kinds of life, not even if it is possible for them to exist, so we can't really discuss them sensibly. What we must really ask, then, is not "Is there life on the moon?" but "Is there life as we know it on the moon?"

As soon as it was discovered that the moon was a world, it was more or less taken for granted that there would be life on it, even intelligent life. The same was true for every other heavenly body that turned out to be a world. The general feeling in earlier times was that the whole purpose of worlds was to carry life, that a world that was empty of life was a wasted world, and that such wasted worlds would not exist. That, however, is just an argument about what we think *ought* to be true. Can we tell whether there's life on the moon without dragging in our personal feelings as to what should or should not be? Remember, until the 1960s, we couldn't go to the moon and actually take a look.

Nor did we have to, for we could tell from a distance. Consider the markings on the moon, the ones that Galileo, with his telescope, found to consist of mountains, craters, and plains. Those markings never changed; the moon was sometimes hidden by earthly clouds, but on a clear night, the smudges were never obscured by clouds on the moon. The moon might be a world like the Earth, but it never had clouds, which seemed to imply that there was no air on the moon within which clouds might form.

This certainly seemed to be true. Every once in a while, as the moon drifted across the sky, it moved in front of a star. If there was an atmosphere about the moon, then as the moon approached the star, that star would begin to shine through the moon's atmosphere and its light would dim slowly until it finally blinked out once it passed behind the solid body of the moon itself. This did not happen. Instead, the star would remain shining at full brightness until it passed behind the solid body of the moon; there was no atmosphere to dim it.

Then again, when we see part of the sunlit side of the moon, we see the boundary between the light and the dark parts. If there was an atmosphere, that boundary would be fuzzy with what we on Earth experience as twilight. On the moon, however, the boundary is sharp; there is no twilight and therefore no atmosphere.

Why is there no atmosphere? The moon has a mass that is smaller than Earth's, and it therefore has a weaker gravitational pull. The surface gravity of the moon is only one sixth that of the Earth, and the moon's gravitational pull is not strong enough to hold an atmosphere. If the moon ever had an atmosphere, it drifted out into space long, long ago.

Nor does the moon have open bodies of water—oceans, lakes, ponds, or rivers. If it did, this water would evaporate under the hot sun, and the moon would not have enough gravity to hold the water vapor either, so even if water once existed on the moon, it would all be gone by now. When Galileo first looked at the moon, he thought the dark areas were seas, which is what they're still called sometimes. Closer looks, however, showed that there were craters and other markings on the "seas" that wouldn't exist there if they were real seas. They seem instead to be lava flows from primordial volcanic action. Since we can easily conclude that the moon lacks air and water, life as we know it is not likely to exist there. The moon has been known to be a dead world, then, even in the 1600s.

Actually, of course, this just means that there are no large and complex forms of life on the moon. It is possible that there are little scraps of air and water here and there in the moon's soil and that very simple forms of life like bacteria can exist there, but surely nothing more.

Yet people would not let go of the notion that worlds must contain life, that a dead world was a wasteful anomaly. In 1835, a British journalist, Richard Adams Locke (1800–1871), wrote a series of articles in the New York *Sun* about the discovery of advanced life-forms on the moon. It was pure fiction, but the public believed it, and the *Sun* became the best-selling newspaper in the world for a brief period. When people want to believe something, they believe it regardless of the evidence. But despite the success of the "moon hoax," the initial studies of the moon through a telescope made it quite clear that dead worlds could and did exist.

42. WHAT CAUSED THE CRATERS ON THE MOON?

The most characteristic markings on the moon are its craters, circular depressions on its surface, surrounded by a mountainous ridge, some of which can be 150 kilometers wide or more. If we think about them, we can easily imagine two different ways in which they might have been formed. The mere fact that we call them *craters* (from a Latin word meaning *cup*, because they seem hollowed out like a cup) reminds us of volcanic craters, and it could be that at an early stage in its history, the moon was very volcanic and that all the craters are the marks of now-dead volcanoes. The other possibility is that the craters were gouged out when large meteorites struck the moon.

People in Galileo's time (and for a couple of centuries afterward) had no experience with meteorite bombardment, but they knew all about volcanoes. Therefore it was taken for granted that the moon's craters were of volcanic origin. To be sure, the craters were much larger than earthly volcanic craters, but the surface gravity on the moon was so much smaller than on Earth that a volcanic eruption on the moon might well kick out far more material than an eruption of the same power on Earth. Even after astronomers became aware of the existence of meteorite impacts, it still didn't seem that the craters could be of meteoric origin. If meteorites struck, they would strike from any direction, and if they hit the moon at a slant, as almost all of them would, it seemed they would gouge out an elliptical crater. Volcanic craters, on the other hand, would always be circular, and the moon's craters were circular.

The first person to seriously question the volcanic origins of the moon's craters was an American geologist, Grove Karl Gilbert (1843–1918), who argued in the 1890s that the moon's craters were altogether different in shape from earthly volcanic craters. Moreover, earthly volcanic craters were almost always on mountain peaks, whereas the lunar craters were at ground level. He could not explain, however, why the craters were round rather than elliptical.

The American astronomer Forest Ray Moulton (1872–1952) finally explained the dilemma in 1929. He pointed out that meteorites would hit the moon at 30 kilometers per second (18.6 miles per second), and the force of such an impact would create something like an explosion in the moon's surface. It would be the explosion, and not the impact itself, that would form the crater, and the explosion, like a volcanic eruption, would always create a circular crater. Meteoric impact was generally accepted to be the cause of crater formation on the moon after that, and scientists now believe that the planetary bodies of the solar system were formed by the joining of smaller pieces; the last pieces to strike left the craters we see today on the moon.

There is no reason why the moon alone should show these bombardment scars, and rocket probes since the 1960s have shown that every airless world has them. Worlds with air can erode away the effects of craters, as can worlds with running water, the activity of life, moving glaciers, flowing lava, and so on. That is why the Earth itself does not seem to have craters, although as I shall explain later, there are ample signs of meteoric bombardment even on this world.

43. HOW WAS THE MOON FORMED?

Earlier I described how the solar system is now believed to have been formed. But that description does not solve all problems, one of which involves our moon. How was it formed?

In general, satellites are much smaller than the planets they circle, so that small planets have no satellites at all, or only very tiny ones. Mercury and Venus have no satellites, and Mars has two, though they are very small, only a few kilometers in diameter.

In 1978, the American astronomer James Christy found that Pluto, the farthest known planet, has a satellite, Charon, that is 10 percent of its own mass. Pluto, however, is a tiny world, smaller than the moon, and Charon is, of course, tinier still.

Jupiter, Saturn, Uranus and Neptune each have numerous satellites, but these planets are far larger than Earth. Some of the satellites of the outer planets are large worlds with diameters of 3,000 to 5,500 kilometers, ranging from a little smaller than the moon to considerably larger. Jupiter has four such large satellites, while Saturn and Neptune have one apiece. Even so, these large satellites are tiny in size and mass, compared with the giant planets they circle.

Earth, however, although a small planet, has a large satellite, one that is far larger in comparison to itself than is true of any of the satellites of the giant planets. The moon has 1.2 percent the mass of the Earth, so that the Earth/moon system might almost be viewed as a double planet.

The first person to tackle the problem of the moon's formation scientifically was the English astronomer George Howard Darwin (1845–1912), who considered the matter of the tides.

I mentioned earlier that as a result of tidal friction, the moon is moving very slowly farther and farther from Earth. This means that yesterday the moon was a little nearer the Earth than it is today, and it was nearer still last year, and even nearer a century ago. In fact, if we look very far back in time, it must have been very close to the Earth, indeed. In that case, Darwin theorized, perhaps the Earth and the moon were once a single body.

The single Earth/moon would have had all the angular momentum that both objects now have separately, so it would have been spinning very rapidly, and it's possible that this rapidly spinning body would have thrown off part of its outermost substance, which would have become the moon. Then tidal friction would have driven it farther away until it assumed its present position.

For a while, this suggestion looked very good. For one thing, the density of the moon is only 3.34 grams per cubic centimeter, so it must be solid rock, without the dense liquid iron core that Earth has. This makes sense because the moon would have been formed only from the outer rocky portion of the Earth, not the inner core.

Then, too, Darwin pointed out that the moon is just about wide enough to fit into the Pacific Ocean, so that it might have broken loose from that part of the Earth. The existence of volcanoes and earthquakes all around the Pacific might be the "scars" that still remain after the forcible rejection of the moon.

Unfortunately, as good as it sounds, Darwin's theory won't work. We know nowadays that the particular shape of the Pacific Ocean changes with time, and that neither it nor the volcanoes and earthquakes at its rim have anything to do with the moon. Besides, if we calculate the total angular momentum of the supposed single Earth/moon body, it turns out that it is only about a quarter as much as would be required to make a portion of the outer crust break loose. For this and other reasons, too, astronomers are now quite certain that Darwin's suggestion that the moon was thrown loose from the Earth is quite wrong.

That would seem to mean that Earth and moon were formed separately from the start, an idea that gives rise to two possibilities. The first is that both the Earth and the moon might have come from the same swirl of dust and gas when all the planets were forming, but for some reason formed a double planet instead of a single one. The second possibility is that they were originally two independent planets formed from two separate swirls. The moon, however, was in an orbit that brought it fairly close to Earth every once in a while, and on one of its close passes Earth's gravity might have captured it.

The thought that Earth and moon formed from the same swirl of dust and gas doesn't look promising, since both worlds would then have to be composed of both rock and metal, and the moon, like Earth, would have to have a metallic core, which it doesn't. On the other hand, if the two worlds formed in two different swirls, one swirl could have been larger and richer in iron, which produced the Earth with its metallic core, while the other swirl could have been smaller and poor in metal, which created the smaller, all-rock moon. But astronomers have not been able to work out a reasonable scenario in which the Earth could have captured so large a body as the moon.

None of the three alternatives that have been suggested for the moon—Darwin's rapid spin, two worlds from one swirl, or two worlds from two swirls and a capture—is a satisfactory explanation for the moon's existence. In fact, one grumpy astronomer said that since all the explanations had failed, the only conclusion one could come to was the moon was not there at all.

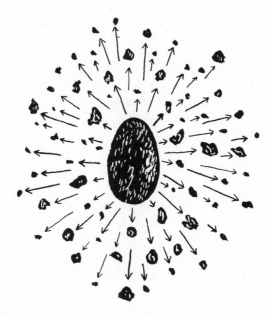

But the moon *is* there, and astronomers had to continue thinking. In 1974, an American astronomer, William K. Hartmann (b. 1939), suggested a fourth alternative. He returned to Darwin's notion of a single Earth/moon body, but didn't rely on the turning of that body to spin off the moon. Instead, he suggested something much more violent: in the first few hundred million years of planetary formation, there must have been considerable chaos; planets were building up out of smaller fragments, and for a while, there were many more sub-planets than exist today and collisions among them were not infrequent. As a result of the collisions, the larger bodies grew at the expense of the smaller bodies until eventually the planets of today had been formed, leaving the rest of space largely clear. In those early days, a second body much like Earth, but with only about 10 percent of its mass, might have slammed into Earth. (This must have happened more than 4 billion years ago, before life had developed on Earth; if it had happened after life had developed, it would have been destroyed by the collision, and life as we know it would have had to develop all over again.) The two objects, each with a metal core, probably would have coalesced, but portions of their rocky outer layers might have erupted into space and formed the moon. This fourth alternative avoids all the difficulties of the first three, without introducing any serious difficulties of its own. At first Hartmann's suggestion was ignored, but by 1984, computer simulation of two sizable bodies colliding made it look as though the idea was a feasible one, and now it is rapidly gaining acceptance.

44. CAN WE REACH THE MOON?

Since we have already done so, the answer is yes, but long before any method existed for going there, imaginative people were already writing about trips to the moon. These were simple fantasies at first, intended to amuse the reader, and often no effort was made to depict the moon realistically. After all, virtually nothing was known of the real moon in ancient times, and it was treated as just one more distant land like India or Ethiopia.

The first tale of a trip to the moon we know of is by a Greek writer, Lucian (c. 120–c. 180). Written in about 165, his story involved a flight to the moon accomplished by his hero through the use of bird wings. Some time later, he wrote another tale in which the hero was carried to the moon by a whirlwind. In 1532, the Italian poet Ludovico Ariosto (1474–1533) wrote an epic poem, *Orlando Furioso,* in which the hero reached the moon by riding the same chariot that had carried off the prophet Elijah in the biblical story. Johannes Kepler had his hero reach the moon in a dream; this story was the first to try to give the moon its real properties, for in it he described its two-week-long daytime and its two-week-long nighttime.

Journeys to the moon became even more popular once Galileo's telescope had shown it to be a real world. In 1638, a book called *Man in the Moone* was published posthumously, having been written by an English writer, Francis Godwin (1562–1633). In it the hero reached the moon by traveling in a vehicle to which large birds were hitched. All of these trips assumed that there was air filling the space between Earth and moon, which seemed a natural assumption at the time; there was air everywhere on Earth, even on the top of mountains. Why shouldn't it stretch upward indefinitely? That this was not so was discovered in 1643, and it came about as follows:

Water can be pumped upward from a depth to a height, but only for about 10 meters (33 feet), and no higher. Galileo was puzzled by this fact, and in 1643, he asked one of his students, the Italian physicist Evangelista Torricelli (1608–1647), to investigate the matter.

To Torricelli, it seemed that what happened during pumping was that some air was sucked out of the cylinder of the pump, which extended below the surface of well water. The air pressing down on the well water seemed to force it up the cylinder once the air was partly removed from within it, and as the pump continued to work, removing more and more air, the water was pushed farther and farther up the cylinder. When the water reached its limit of 10 meters (33 feet), it seemed as if the column of water had produced as much downward pressure as the air itself could exert; the two pressures, that of the column of air pressing down on the surface of the well water and that of the column of water inside the pump cylinder, seemed to be equal and the water could rise no farther.

To check this theory, Torricelli made use of mercury, a liquid with a density nearly 13½ times that of water, so that a column of mercury ought to exert 13½ times as much pressure as a column of water of the same length. If air pressure could uphold 10 meters (33 feet) of water, it ought to uphold only 0.76 meters (30 inches) of mercury. Torricelli filled a 1.2-meter (47 inches) length of glass tubing with mercury. He stoppered the opening and upended it, stoppered end down, into a large dish of mercury. When the tube was unstoppered, the mercury began to empty out of the tube, but not altogether; a column of mercury 0.76 meters (30 inches) high remained in the tube. The mercury clearly pressed down with a pressure equal to the air pressure (which was the weight of a column of air the same width as the mercury column and stretching to the top of the atmosphere).

This showed, first, that air had weight and therefore mass. It was not a massless vapor, but a material object, and even though it was thinly spread out matter, it was still matter, nonetheless. Second, the fact that a column of air held up only 0.76 meters (30 inches) of mercury showed that the pressure was finite and that, therefore, so was the height, which meant that the weight of a volume of air could be accurately measured and its density determined. Air has a density of 0.0013 grams per cubic centimeter, so that it is only $1/77$ as dense as water. Thus, if the density of the atmosphere remains constant throughout, it must be only 8 kilometers (5 miles) high.

As a matter of fact, the density of the atmosphere is not the same throughout. The upper layers weigh down and compress the lower layers, and since air is much more easily compressed than rock is, the lower layers are much denser than the upper layers. This fact was proved by the French physicist Blaise Pascal (1623–

1662), who in 1648 sent his brother-in-law up a mountainside carrying tubes of mercury. If the air were evenly dense all the way up, then by the time the height of a mile had been reached, the column of mercury should be only four fifths what it was at sea level (0.61 meters, or 24 inches). The column of mercury did become lower, but not quite that fast. As one went up, the air was rarefied and spread out more, so that the atmosphere extended higher than expected.

Just the same, even allowing for a wider atmosphere, by the time you moved up 160 kilometers (100 miles), so little air was left above you that it might as well be ignored. It thus turned out that if you traveled from the Earth to the moon, you would be traveling through vacuum for more than 99.95 percent of the distance. In fact, except for regions in the immediate vicinity of a large body, space is a vacuum containing only minuscule traces of matter.

If we stop to think about it, we can see that this must be true. If air filled the universe, as was taken for granted in the days before Torricelli, then the moon and other bodies would be traveling through air and would have to lose energy steadily as they pushed the air to one side. The moon's motion would slow down until it very gradually fell to Earth, while the Earth's motion would decrease, too, so that it would very gradually fall into the sun. In fact, the only reason that objects in space can maintain their orbits indefinitely is that they are traveling through a vacuum and lose virtually no energy in so doing.

The fact that outer space is a vacuum presents certain difficulties as far as travel to the moon is concerned. The moon can't be reached by flying birds or waterspouts, and we can't really rely on magic chariots or dreams. The only practical method known for traveling through a vacuum makes use of the rocket principle. In 1687, Newton pointed out as one of the laws of motion that if part of the mass of an object is hurled in one direction, the rest of the mass must move in the opposite direction (the law of action and reaction). Therefore if a vessel contains a quantity of material that can be converted into hot gases, and if those hot gases are allowed to escape downward at high speed through a narrow opening, the vessel itself must move upward, and if it can attain a high enough speed, it would be able to leave the Earth permanently.

In 1650, a book called *Voyage to the Moon* by the French writer Cyrano de Bergerac (1619–1655) mentioned seven different ways in which one could reach the moon. Six of them were simply fantasy and could not possibly work. The seventh, however, was rocketry—thirty-seven years before Newton had established the principle. In 1926, an American physicist, Robert Hutchings Goddard (1882–1945), built and launched the first liquid-fuel rocket of the modern type. It was a tiny thing, but it showed the way, and on July 16, 1969, the American astronaut Neil Alden Armstrong (b. 1930) became the first human being to set foot on the moon.

Trips to the moon quickly demonstrated that the moon was indeed an airless, waterless, and lifeless world. There was no sign that even the simplest microscopic life was present on the moon, or had ever been present.

Moon rocks were then brought back to Earth for analysis. Since the moon is smaller than the Earth and has less heat at the center, it is less turbulent and has less volcanic action. Therefore its rocks might well have existed on the moon's surface, unchanged, for longer than any on Earth. In fact, some moon rocks turned out to be 4.2 billion years old, half a billion years older than the oldest surviving rocks on Earth seem to be.

45. WHAT ARE METEORITES?

Anyone watching the sky on a dark night is bound to see, on occasion, something looking like a star streak across the sky and disappear. It looks exactly as though a star has slipped from its place in the sky and slid downward along its curve. In fact, the phenomenon is popularly referred to as a "falling star" or a "shooting star."

Even the ancient Greeks, however, noted that no matter how many shooting stars were seen, none of the familiar fixed stars were ever missing from the sky. Whatever the shooting stars were, then, they were not really stars. The Greeks evaded the necessity of explaining them by simply calling them *meteors* (from Greek words meaning *objects in the air,* which is just about the equivalent of calling them unidentified flying objects, or UFOs).

We now know that meteors are bits of matter the size of pinheads or less. Nearby space is littered with these particles (you might say space is "dusty"), and when one of these particles happens to come near the Earth, it compresses the air in front of itself. The air compression raises the particle's temperature until it glows and vaporizes and breaks down into even tinier dust particles. Such dust doesn't affect us adversely, and it is even extremely useful for acting as nuclei around which water droplets form. The dust thus contributes to rainfall, which is essential to life on land. (The question of where all this dust comes from will be taken up later in the book.)

There are, however, bits of debris that strike the Earth that are considerably larger than pinheads. Some are so large that they survive the passage through the atmosphere and hit the Earth's surface as relatively large objects. Such larger pieces of debris traveling through space are known as *meteoroids,* and the pieces that hit the ground are *meteorites.* About 10 percent of the meteorites are nickel-iron in composition, which, as I explained earlier, is what gave scientists the first notion that the Earth's core might be nickel-iron.

Ancient people occasionally came across nickel-iron meteorites at a time when they had not yet learned how to smelt iron from iron ore. This free gift of a particularly tough form of iron (with nickel added) was of enormous value, for it made better, harder, tougher, sharper tools than anything else available to them. Thus, the *Iliad* describes a lump of iron given as a prize at the funeral games of Patroculus that was undoubtedly a meteorite. There are no such meteorites to be found in the developed regions of the Earth, however, for all have been scavenged.

Sometimes a meteorite may actually be seen to fall. The

second-century Greek astronomer Hipparchus is supposed to have been told of someone witnessing the fall of a meteorite, an event that may have been viewed as a sign from heaven. A fallen meteorite was worshiped in the Temple of Artemis in Ephesus in ancient times, and the black stone in the shrine of Kaaba in Mecca is probably a fallen meteorite, too.

The tales of stones falling from the sky were disbelieved by astronomers in early modern times. An American chemistry professor, Benjamin Silliman (1779–1864), together with a colleague, reported having witnessed such a fall in 1807, but Thomas Jefferson (1743–1826), then president of the United States and an accomplished scholar, remarked that it was easier to believe that two Yankee professors would lie than that rocks would fall from heaven.

(It is easy to sneer at scientists for being overly skeptical, but it is far safer to be skeptical and to allow time and accumulating evidence to prove the truth of an unpopular view than to be too eager to accept new ideas and waste scientific effort on hare-brained pursuits in every direction.)

There were scientists who took up a minority view. The German physicist Ernst F. F. Chladni (1756–1827) published a book in 1794 in which he argued that stones *did* fall from heaven, and he even collected objects that were reported to have so fallen. A new report of falls in France in 1803 resulted in a careful investigation by a French physicist, Jean Baptiste Biot (1774–1862), and his report convinced the scientific world, at last, that meteorites really existed.

They have been carefully studied ever since, for until 1969 they were the only samples of non-earthly matter that were available. For another thing, they were very small and had existed indefinitely in the vacuum of outer space, so it was very likely they had not changed or been disturbed since they were first formed.

Some meteorites turned out to be 4.6 billion years old, older than anything we could find undisturbed on either the Earth or the moon, and it is this 4.6-billion-year age that was used to give the birthdate of the solar system, including the sun, the moon, and the Earth.

46. MIGHT METEORITES BE DANGEROUS TO LIFE AND PROPERTY?

Of course. It doesn't take much thought to determine that if the Earth is bombarded randomly with bits of rock and metal, then sooner or later, one of these missiles will hit someone. So far, although meteorites are known to have struck houses and even cars, there are no records of anyone having been killed by one. It stands to reason, though, that it is only a matter of time before someone is.

The Earth is a huge target, and it is much more likely that a meteorite will hit the ocean or a desert or sparsely occupied woodland or farmland than a human being or even a city. Still, the number of people is increasing steadily, cities are spreading outward, and the Earth is becoming fuller and fuller of human-made structures, so the target is growing larger, and eventually a falling meteorite may cause some sort of tragedy.

Of course, the larger the meteorite, the more damage it can do, but large meteorites are far rarer than small ones. The largest strike of historic times that we know of took place in 1908, when a sizable object struck in central Siberia and knocked down every tree in the forest for about 32 kilometers (20 miles). It wiped out a herd of deer, but the area was uninhabited and no human lives were lost. Perhaps 25,000 years ago, a meteorite that was larger still struck what is now Arizona and gouged out a crater that is 0.8 kilometers (0.5 miles) wide. Thanks to the fact that it is in a desert area and has not had to suffer the effects of water and human activity to any great degree, the crater is still visible. Had such a meteorite struck now and hit a city, the entire city would have been wiped out in a flash.

There are signs of even larger strikes a few million years ago, leaving still larger craters that have been wiped out by the action of wind, water, and plant growth, but whose existence can still be determined. Before we can discuss the largest strikes, however, we must take up another subject.

47. WHAT ARE ASTEROIDS?

One of the reasons scientists in the 1700s found it so difficult to accept the existence of meteorites is that there was no knowledge of any small bodies in the solar system. There seemed to be only the planets and their satellites (and also the mysterious comets which I will talk about later).

A change of mind slowly began with a German astronomer, Johann Daniel Titius (1729–1796). In 1766, he worked up a formula to show the relationship of the mean distances of the planets from the sun. It gave the following numbers: 4, 7, 10, 16, 28, 52, 100, 196, 388, and so on. Suppose we allow Earth's distance from the sun to be represented by 10. In that case, the distance of Mercury from the sun is about 3.88, that of Venus is 7.23, that of Mars is 15.23, that of Jupiter is 52.0, and that of Saturn is 95.5. In 1772, another and better known German astronomer, Johann Elert Bode (1747–1826), publicized the number series, which came, rather unjustly, to be known as *Bode's law*.

Eventually, astronomers began to notice that there was no planet in position 28 in Bode's law. Should there be a planet there? If so, why had it never been seen? It would be at only twice the distance of Mars from Earth and only two fifths the distance of

Jupiter from Earth. Even if this position 28 planet were no larger than Mars (which has only a little over half the diameter of the Earth) it still ought to be easily visible. The only possible reason a planet could be in position 28 and not be seen would be that it was much smaller than Mars.

A German astronomer, Heinrich W. M. Olbers (1758–1840), began to organize an astronomical project in the 1790s whereby each of a number of astronomers would take separate parts of the sky and search it carefully for a possible planet with an orbit between those of Mars and Jupiter. Before it could go into action, however, an Italian astronomer, Giuseppe Piazzi (1746–1826), made the discovery on January 1, 1801, the first day of the nineteenth century. He wasn't looking for it; he just happened to come across a "star" that changed position from evening to evening so that it could not be an ordinary star. From its speed of motion, it seemed to be the missing planet between Mars and Jupiter. Since Piazzi was a Sicilian, he named the new planet Ceres, after the goddess of agriculture who had been worshiped in ancient Sicily.

Ceres is a small planet, for it is only about 1,000 kilometers (620 miles) in diameter, less than half the diameter of the moon.

But Olbers found it difficult to believe that that was all there was in the space between Mars and Jupiter, and continued the search as planned. In the course of the next few years, three more such bodies, even smaller than Ceres, were discovered between Mars and Jupiter, and they were named Pallas, Vesta, and Juno.

The German-British astronomer William Herschel (1738–1822) pointed out that the new bodies were so small that they looked merely like points of light, as the stars did, even through the telescope, and did not show visible orbs like the larger planets. He suggested, therefore, that they be called *asteroids* (*starlike*), and the name has stuck.

Since Piazzi's discovery, asteroids have been discovered in enormous numbers. Over three thousand are known, and, undoubtedly, there are many thousands of others in the space between Mars and Jupiter. Ceres remains the largest, making up nearly 10 percent of the mass of all the asteroids. The space between the orbits of Mars and Jupiter is therefore referred to as the *asteroid belt,* and it is a region that must be faintly reminiscent of what the solar system must have been like before the planets were fully formed.

Why are the asteroids there? Olbers was the first to suggest they were the remains of an exploded planet. This is an attractive idea, but we don't know how or why a planet should explode. Astronomers nowadays think that the material in the asteroid belt simply couldn't ever condense into a planet. Jupiter, a giant planet, might have swept up so much of the material of the asteroid belt that what was left over could not form a sizable planet. Moreover, Jupiter's gravitational pull might well have kept the asteroids from coalescing.

48. ARE THERE ASTEROIDS ONLY IN THE ASTEROID BELT?

There are thousands upon thousands of asteroids, most of them very small—just jagged hills on the loose. Even if they were all in the asteroid belt to begin with, they wouldn't necessarily stay there. As the asteroids orbit the sun, they are affected by the gravitational pulls of the other planets, particularly that of giant Jupiter. Some may drift out past the orbit of Jupiter into the outer solar system, while others may drift inward past the orbit of Mars into the inner solar system. The farther out asteroids go, the harder they are to see and study, so we don't know much about the more distant ones. On the other hand, those that come in closer than Mars are easier to see and study and, as you can well understand, more dangerous.

In 1898, a German astronomer, Gustav Witt, discovered an

asteroid whose orbit carried it inside that of Mars, and he named it Eros. (Asteroids are usually given feminine names, but those with peculiar orbits are given masculine ones.) When Eros and Earth are at just the proper places in their respective orbits, they are separated by only 22.5 million kilometers (14 million miles)— only a little over half the distance of the closest approach of the planet Venus. Eros therefore approached more closely than any astronomical object then known, except for the moon. In 1931, it passed within 26 million kilometers (16 million miles) of Earth.

Such a distance is safe enough; it is extremely unlikely that its orbit will ever change to the point where it will collide with us, and that's a good thing, for its average diameter is about 16 kilometers (10 miles). A collision with Eros would not necessarily damage the Earth itself by much, but it would be catastrophic to life on Earth.

The trouble is that Eros isn't the only asteroid of its kind. In the years since 1898, a number of asteroids (mostly only 1 or 2 kilometers in diameter) have been discovered that can approach Earth even more closely than Eros does. At least fifty of these "Earth grazers" are now known, and a few additional ones are discovered each year.

The meteorites I discussed earlier are tiny examples of wandering asteroids. They don't do much damage, but sooner or later, one of the larger Earth grazers is bound to hit the Earth. In fact, according to some estimates, there is such a catastrophic collision an average of once every 100 million years. If this is so, then over thirty such collisions might have taken place while there was life on Earth. There might have been five or six such collisions while complex forms of life existed on land and sea. Are there any traces of these collisions?

About 65 million years ago, a change of some kind took place on the Earth that caused the dinosaurs, together with other types of plants and animals, large and small, to vanish suddenly from the face of the Earth. Until 1980, no one was quite sure what had happened; there were numerous theories, but none carried conviction. In 1980, however, the American scientist Walter Alvarez was analyzing, with great delicacy, rock layers that were 65 million years old. He found them twenty-five times as rich in the rare metal iridium as were the layers that were just a little bit older or a little bit younger. Something had doused the rock with iridium just as the dinosaurs disappeared. Nor was it just in that one place where Alvarez happened to be working, for a similar enrichment of iridium was found in rocks of that age all over the world.

What happened? Alvarez argued that iridium is much more common in meteorites than in Earth's crust. (On Earth, most of the iridium was concentrated in the iron core.) It seemed, then, that there had been a particularly large strike 65 million years ago that had vaporized the meteorite, together with cubic miles of the Earth's crust, with the enormous heat of the collision. Vast quantities of dust were flung up into the upper atmosphere, cutting off the light of the sun for a long period and producing an artificially prolonged winter that wiped out many forms of life. It also might have caused earthquakes, volcanic eruptions, floods, vast forest fires, and so on. Most of life, especially the large animals, vanished. Smaller forms of life, or just lucky ones, managed to survive and begin over again.

There are signs that this has happened periodically throughout Earth's history. Every once in a while, there is a vast extinction that wipes out much of life. It is possible that this is an important part of evolution, since it gives new forms of life a chance to develop and expand. For instance, mammals existed tens of millions of years before the last "great dying," but could not compete with the giant dinosaurs; they remained small and unimportant. It was only after the meteor strike had wiped out the dinosaurs that the small mammals had a chance to undergo explosive evolution and develop the many advanced forms that exist today—including us.

If there is another such collision in the future, and if we haven't managed to wipe ourselves out before then, all human life could be destroyed, leaving the planet, perhaps, to some other form of life to write a fresh page. So far, at least, no collision has been terrible enough to sterilize the Earth completely, but we can't be sure that so dreadful a catastrophe is entirely impossible.

49. WHAT ARE COMETS?

There is another kind of object, in addition to asteroids and meteorites, that can approach the Earth—the *comet*. It might have been a collision with a comet, rather than an asteroid, that caused the great extinction of the dinosaurs 65 million years ago. It might also have been a sliver of a comet, rather than a meteorite, that caused the explosion in central Siberia in 1908. What, then, is a comet?

Comets are much more apparent than meteorites. They are not merely streaks of light that come and go in a matter of seconds, but hazy objects, sometimes quite large, that remain in the sky for weeks on end. In the past, they were objects of fear. Whereas stars and planets followed well-worn paths and their motions were predictable, comets seemed temporary objects that came out of nowhere, drifted across the sky from night to night, and finally disappeared. If people believed that the planets formed patterns as they moved across the sky, patterns that predicted the future, they were bound to assume that a comet was a kind of one-shot message, sent as a warning by an angry deity.

The thought that a comet was a harbinger of bad news, rather than good news, seemed to be confirmed by the comet's appearance. It consisted of a foggy ball of light with a long, luminous tail extending to one side. People with good imaginations thought its shape represented the head of a mourning woman shrieking her lamentations across the sky, with her unbound hair flowing behind her (indeed, the word *comet* is from the Greek word for *hair*); others saw it as a sword. Either way it meant death and disaster, and you could argue that this was true because every time a comet appeared, catastrophe struck. Of course, catastrophe also struck every time a comet did *not* appear, but somehow people didn't notice that.

Some ancients tried to be rational about these objects. Since Aristotle believed that the heavens were perfect and unchanging, there was no room in them for such changeable, temporary things as comets. He therefore thought that they were merely flaming gases in the Earth's upper atmosphere, rather like the will-o'-the-wisps that sometimes flicker over marshy areas. This theory was incorrect, but it was at least a sensible interpretation. It did not, however, persuade the general public to cease its foolish fears. (Even in the twentieth century, there are people who are frightened by comets, just as there are still people who actually believe the Earth is flat. In the twentieth century, however, there has been no really spectacular comet since 1910, so that the fears have not had a chance to blossom.)

The first scientist to study a comet dispassionately was the German astronomer Regiomontanus (1436–1476), who observed a comet that appeared in the sky in 1473 and recorded its position from night to night. In 1540, a German astronomer, Petrus Apianus (1495–1552), published a book in which he described five different comets. In it, he pointed out that in every case, the tail pointed away from the sun. This was the first scientific observation made about any comet other than its position in the sky.

In 1577, Tycho Brahe tried to determine the parallax of a bright comet that appeared that year, but was unable to do so; its parallax was not large enough to be measured, though the moon's was. This meant that the comet was farther away than the moon. Aristotle was wrong, then; comets were not to be found in Earth's atmosphere, but farther out in space.

Once Newton had worked out his law of gravitation, it seemed natural to suppose that it applied to comets as well as to everything else in space. Comets ought to be held by the gravitational pull of the sun, and they ought to revolve about the sun. The only trouble was that whereas ordinary planets traveled in ellipses that were almost circles, comets seemed to have very elongated orbits. Perhaps they entered the solar system only once, zoomed around the sun, and then left never to return.

The English scientist Edmund Halley (1656–1742), a friend of Newton's, tackled the problem. He studied earlier records of comets and found that the comets of 1456, 1531, and 1607 followed the same path across the sky that had been followed by the comet of 1682, which he himself had observed. It occurred to him that this was the same comet returning to that portion of its very elongated orbit that carried it near the Earth and the sun every seventy-five or seventy-six years.

Halley stated that this same comet would return in 1758. He did not live to see this, but it returned almost as predicted, in early 1759, and that particular comet has been known as *Halley's comet*, or *Comet Halley*, ever since. Its most recent appearance in Earth's sky was in 1986, but it did not pass particularly close to Earth, and put on a poor show. Halley's discovery removed much of the mystery from comets, and for several decades it became a fad among astronomers to discover new comets and work out their orbits.

50. WHY DO COMETS LOOK FUZZY?

Even after comets were revealed as ordinary members of the solar system, subject to the law of gravitation, they remained mysterious. Other objects in the solar system were solid bodies with sharp edges and had no tails, while comets were fuzzy and had tails. Most small bodies in the solar system, such as Mercury, the moon, the asteroids, and most satellites, are lumps of solid matter without an atmosphere, and they naturally have sharp edges, just as any rock or lump of metal on Earth would. The giant planets have atmospheres, as do Earth, Venus, Mars, and even a couple of the larger satellites. These gaseous envelopes are held firmly to the object by gravitational forces and either don't interfere with the sharp edge of the solid body of the planet below or form cloud layers that themselves present a sharp boundary.

Comets are unlike any of these bodies in that they have a

different chemical structure. (You might immediately ask at this point how astronomers can tell the chemical structure of a distant object, but this is a subject that will concern us later.) Though comets are small objects like asteroids, they are not composed of rock and metal, but of substances that are *volatile* (easily melted), which would ordinarily be liquids or gases on Earth, but are frozen into solids at low temperature. The most common of such volatiles is water itself, both on Earth and in the comets. In the comets, it exists as solid ice. Other volatiles, like ammonia and cyanogen, can freeze into solids resembling ice in appearance; they are all lumped together as *ices*.

Comets are made up of ices, with grains of rock and metal mixed in, and possibly a rocky core. This structure was worked out in 1949 by the American astronomer Fred Whipple (b. 1911), who referred to comets as "dirty snowballs." As long as these objects remain far from the sun, they are frozen solid and have sharp edges just as asteroids do, though they are then too far away for us to be able to see and study them. As they approach the sun, however, the sun's warmth evaporates some of the ice and frees some of the rocky dust it contains. The solid nucleus of the comet is then surrounded by a cloud of gas and dust. The dust particles reflect the sunlight, surrounding the comet with a luminous fog, the *coma*, which gives it a fuzzy appearance.

There are always electrically charged particles emerging from the sun in all directions, and this is called the *solar wind*. It is a very feeble wind, indeed, but it is strong enough to drive the cloud of dust and gas away from the comet, so that a long, luminous tail is formed that always points away from the sun.

51. WHAT HAPPENS TO COMETS?

A comet isn't a permanent object in the sense that the Earth or an asteroid is. When a comet travels around the sun and some of it is evaporated, the evaporated portion never returns. The wonder is, perhaps, that the comet doesn't evaporate entirely and disappear in the fiery embrace of the sun, and, in fact, it would do just that if it lingered too long in the sun's vicinity. Instead, it races past it and away before very much of it evaporates.

As the ices vaporize, some of the rocky dust is left behind, forming a crust on the comet's surface. Probes into the vicinity of Comet Halley in 1986 showed that its surface was black with rocky dust. Such a rocky crust acts as a bit of an insulator, cutting down on evaporation.

Nevertheless, some of the comet's material *is* lost at each near passage of the sun, and comets experience only a temporary existence. Even large ones fade away after some hundreds or perhaps thousands of passages near the sun. Astronomers have watched some small comets fall into the sun and vanish forever and others break up and disappear. Some comets leave behind a rocky core that is all but indistinguishable from an asteroid. Others leave behind only ghosts of themselves. While the gases evaporate and spread out through space, the dust that has been freed by the evaporation continues to move in the cometary orbit, spreading out along it and thinning, but staying denser in the spot where the comet used to be.

On November 13, 1833, the Earth and the main dust cloud of a dead comet collided. It didn't do the Earth any harm; indeed, it presented a glorious sight, for the skies of New England turned into a fireworks show. Uncounted numbers of dust particles streaked through the atmosphere, glowing like the falling of luminous snowflakes that, however, never reached the ground. It seemed to awed onlookers that all the stars in the sky were falling, and since the Book of Revelation states that on the Day of Judgment the stars would fall from the sky, there must have been many who thought the end of the world was at hand. The next day, however, the sun rose as usual, and the next night all the stars were still in the sky.

There are several times in the year when the number of meteorites is greater than usual, but the 1833 display has never been repeated, though it did stimulate the further study of meteorites.

52. WHERE DO COMETS COME FROM?

If comets are short-lived, if they tend to break up and disappear and leave behind only a rocky core or some dust, why are they still around? Why haven't they all disappeared in the course of the 4.6-billion-year lifetime of the solar system?

If we think about it, there seem to be only two possible answers: either new comets are being formed as fast as the old ones disappear, or there are so many comets that even in 4.6 billion years they have not yet all been used up. The first possibility does not seem very likely, because astronomers can think of no way in which comets are still being formed.

That leaves the second alternative. In 1950, a Dutch astronomer, Jan Hendrik Oort (b. 1900), suggested that when the solar system formed, the outermost reaches of the vast dust and gas cloud that condensed to form it were not sufficiently held by the gravitational pull of the far removed interior to become part of the condensation. While the inner regions condensed, the outermost regions stayed where they were and underwent minor condensations into at least 100 billion lumps of icy materials. This cloud, located far beyond the planetary stem but still in the gravitational grip of the very distant sun, is called the *Oort cloud* in the astronomer's honor. No one has ever seen the cloud or detected it in any way, but it is the only means, so far, by which the existence of comets at the present time can be accounted for.

Apparently, the comets in this vast cloud are moving in constant but slow motion in mighty orbits about the sun, completing each turn only in many millions of years. Every once in a while, though, either because of a collision between two comets or because of the pull of the nearer stars, a comet's motion changes. It can speed up, in which case its orbit expands still farther from the sun or takes it out of the solar system forever. It can also slow down and drop toward the inner planets of the solar system and pass near the sun, in which case it might appear in Earth's sky as a magnificent spectacle, and since it keeps its new orbit (except when this is changed by planetary attraction), it eventually evaporates and dies.

Oort estimates that in the lifetime of the solar system, one fifth of all the comets have either been driven out of the solar system or made to drop into the inner solar system to evaporate. That, however, still leaves four fifths of the original supply to serve as an ongoing reservoir of comets.

53. HOW FAR IS THE SUN?

I mentioned the distances of the planets in connection with the discovery of the asteroids; by the time of that discovery, those distances were known. However, for over eighteen centuries after Hipparchus had determined the distance of the moon, that remained the *only* distance known, for there was simply no way of measuring the parallax of any object farther away.

The Greek astronomer Aristarchus (c. 310–c. 230 B.C.), as I explained earlier in the book, made an attempt to determine the distance of the sun without the use of parallax. His method, in 270 B.C., was perfectly correct in theory, but he had no way of determining angles in the sky accurately, and his guesses were way off. He ended with the decision that the sun was about 8 million kilometers (5 million miles) away from the Earth and that the sun had seven times the diameter of the Earth.

This was a great underestimate, but it was sufficient to cause Aristarchus to think that the Earth went around the sun, rather than vice versa. But no one took either his figures or his conclusion seriously.

By the 1600s, however, after the discovery of the telescope, it was possible to determine the position of a heavenly object far more accurately (especially after cross hairs were placed in front of the lens). This meant that a small shift in an object's position, a tiny parallax, one that could not be measured by the eye alone, could be measured with the telescope. But it was not necessary to measure the parallax of the sun in order to determine its distance. That would be a hard job, indeed, for positioning its glowing edge would be almost impossible, especially since there are no stars visible in the sky when it is shining, against which its position could be measured.

Instead, the parallax could be determined for any planet. Thanks to Kepler's model of the solar system, which is still considered correct today, the distance of any planet at any particular position in its path about the sun could be used to calculate the distances of all the planets from each other, from the sun, and from

us. Of course, we needn't judge the worth of a world by size alone, but the translation into tininess wasn't easy to take.

Furthermore, we could not argue that Jupiter and Saturn might be large but insubstantial. Each had satellites that orbited their planets at known distances and periods. The faster a satellite at a given distance travels about a planet, the stronger the planet's gravitational pull and, consequently, the greater its mass. By comparing the motion of satellites about Jupiter and Saturn with that of the moon around the Earth, it turns out that Jupiter has a mass 317.9 times that of the Earth, and Saturn one of 95.2 times that of the Earth.

Even so, the masses of Jupiter and Saturn are not as large as one would expect from their sizes. If we divide the mass of each by its volume, it turns out that the average density of Jupiter is 1.33 grams per cubic centimeter, which is less than a quarter of the density of Earth. Saturn is even less dense, 0.71 grams per cubic centimeter, only about an eighth as dense as Earth and actually less dense than water. This means that the compositions of Jupiter and Saturn must be widely different from that of Earth, a point we'll return to later.

55. ARE THERE PLANETS THE ANCIENTS DIDN'T KNOW?

The asteroids, which I have already described, may be tiny, but they are planets circling the sun, and no one knew of their existence before 1801. We might recast the question, then, and ask are there any *large* planets the ancients didn't know?

Until the late 1700s, this was another question it would have seemed that no sensible person would ask. The seven "wandering stars," the sun, the moon, Mercury, Venus, Mars, Jupiter, and Saturn, were all known to the ancient Sumerians by 3000 B.C. For the next 4,700 years no further objects had been seen to wander among the stars (with the exception of comets). How could there be any planets that were still undiscovered? Since all of the known planets were bright and unmistakable, surely any others would be the same and would have been easily found. So one could only conclude that none were there.

Yet the planets are not gleaming objects that give off their own light, as had been widely believed since the time of the Sumerians. First the moon, from its phases, was found to be a dark body by the ancient Greeks. Later the telescope revealed that Mercury and Venus exhibited phases, too, and were therefore dark bodies. Thus it was assumed that all of the planets were dark bodies that shone only by light reflected from the sun.

In that case, the farther a planet was from the sun and the smaller it was, the less sunlight it would receive, the less it would reflect, and the dimmer it would appear in the sky. If there were other planets beyond Saturn and they were considerably smaller than Saturn, they might be so faint that they'd probably be overlooked by astronomers who expected that all planets were bright. Besides, the farther a planet is from the sun, the more slowly it moves along its path, so that its wandering against the background of the stars might disguise it still further.

All this is perfectly clear in hindsight, but astronomers, even with their telescopes, had gotten so used to thinking that all planets were bright that it never occurred to them to look for a dim one, and, in fact, they disregarded the possibility.

When a new planet was finally discovered in 1781, it was found by accident. William Herschel (who first suggested the term *asteroid*) was a musician by profession, but became interested in astronomy as a hobby. He tried to buy a telescope, found that those he could afford weren't much good, and constructed his own, which turned out to be better than any others then in existence. With this homemade telescope, he came across an object in the sky that looked like a small disk of light, the way planets did. It didn't occur to him at first that it might be a planet, and he assumed it was a comet. But comets are fuzzy, and this disk had sharp edges, and the new object moved more slowly than Saturn against the background of the stars, which indicated that it was even farther away from the sun than Saturn was. In fact, it was a new planet, and it was named Uranus. It is twice as far from the sun as Saturn is—2.87 billion kilometers (1.78 billion miles)—and so dim that it is just barely visible to the naked eye.

Since then, two planets have been discovered that are farther from the sun than even Uranus is—that are so far away, in fact, that they are too dim to be seen with the eye alone and couldn't have been discovered under any circumstances before the invention of the telescope. The planet just beyond Uranus was discovered in 1846 and was named Neptune, and a small planet beyond Neptune was discovered in 1930 and was named Pluto. The extreme stretch across this last planet's orbit is almost 12 billion kilometers (7.3 billion miles), so that, compared with the days before Herschel, when Saturn was thought to be the farthest planet, the new planets have almost quadrupled the known width of the planetary system.

Uranus and Neptune are giant planets, though not as large as Jupiter or Saturn, and they both have diameters of about 50,000 kilometers (31,000 miles), more than 3½ times that of Earth. Uranus has a mass that's about 15 times greater than Earth's; Neptune's is about 17 times greater. Their densities are roughly equal to that of Jupiter. It turns out, then, that Earth is only the sixth largest object in the solar system as it is presently known; the sun and the four planets Jupiter, Saturn, Uranus, and Neptune are all considerably larger. Astronomers are still searching for another sizable planet beyond Neptune (Pluto is so small it can scarcely be counted), but they have not yet found one.

56. IN WHAT WAYS ARE THE GIANT PLANETS DIFFERENT?

The four large planets of the outer solar system are different in many ways from the Earth and the more familiar worlds of the inner solar system. There is their low density, for instance, which means that they are constructed of entirely different materials from the Earth, as we shall see later. They all have large and deep atmospheres with permanent cloud layers that we see as the *surface* of those planets (we don't see a solid surface).

Jupiter, being nearest to the sun, has the most energy available to it so that its atmosphere is churning with enormous storms. The most important is an apparently permanent tornado, larger than the Earth, called the *Great Red Spot* because of its color. It was first noted by the English scientist Robert Hooke (1635–1703) in 1664.

Saturn and Uranus are quieter than Jupiter, but Neptune, the farthest of the four giants, was revealed in 1989 by the probe *Voyager 2* to have winds as great as those of Jupiter. Scientists are not sure why that should be. It also has a *Great Dark Spot* similar in shape and location to the spot on Jupiter. (Jupiter is the real giant, by the way, since fully 70 percent of the total mass of the solar system outside the sun is to be found there.)

The giant planets all have numerous satellites. Most of them are fairly small, but Jupiter has four, which were discovered by Galileo in 1610, that are the size of the moon or larger. Saturn has one, Titan, which was discovered in 1655 by Huygens. Neptune has one, Triton, which was discovered in 1846 by the British astronomer William Lassell (1799–1880).

Of the four, Uranus has the strangest rotation. All the planets have axes that are more or less tilted to the plane in which they circle the sun. The Earth, for instance, is tilted about one quarter of the way over, as are Saturn and Neptune. Jupiter's axis is almost, but not quite, untilted. Uranus, however, has its axis tilted so far that it seems to be rotating on its side. It makes one revolution about the sun in eighty-four years, so that at one point in its orbit its north pole is facing almost directly toward the sun and forty-two years later its south pole is facing the sun. Presumably, the planets were knocked this way and that as they formed from planetesimals, and by sheer chance, they were knocked more in one direction than another so that their axes of rotation tilted. In the case of Uranus there must have been an unusually unbalanced coming together of the final planetesimals, which caused it to be knocked onto its side.

The prize of the giant planets, however, is Saturn. When Galileo first turned his small telescope on it, it was the farthest known planet, and he couldn't make it out very well. Even so, it seemed to him it had a bulge on either side. Could it be a triple planet? That didn't seem to make sense, and by 1612, he stopped trying to observe it. In 1614, a German astronomer, Christoph Scheiner (1575–1650), viewing Saturn through his telescope, thought that there wasn't a bulge on either side, but a crescent, as though Saturn had a teacup handle on each side. The mystery wasn't solved till 1655, when Huygens (the inventor of the pendulum clock) looked at Saturn and found a flat ring that circled the planet's equator without touching it. In 1675, Cassini (who was the first to obtain the parallax of Mars) noticed a dark marking that seemed to divide the ring into two rings, one inside the other. The marking is called *Cassini's division* to this day.

The rings are bright, brighter than the globe of Saturn, and they are enormous, too. They turn Saturn into what is, in the opinion of many observers, the most startling and beautiful sight one can see through the telescope. From the outer edge of the rings on one side of Saturn to the outer edge on the other, they stretch across a distance of 272,000 kilometers (168,600 miles). It would take 21½ globes the size of the Earth to stretch across the rings from end to end. They are more than twice the width of the planet, though, of course, they are thin rings (like a playing record) and add little to Saturn's mass.

But what are the rings? Are they solid disks of matter? In 1859, the British mathematician James Clerk Maxwell (1831–1879) showed that if the rings were solid, Saturn's tidal effect, stretching them powerfully toward and away from Saturn, would break them up. He concluded that the rings consisted of individual particles and looked solid only because of their great distance, just as the seashore looks like a solid stretch of land until you get close enough to see that it is made of individual sand grains.

Then why are the rings there?

In 1850, a French astronomer, Édouard Roche (1820–1883), tried to work out what would happen if the moon were somehow circling nearer the Earth. He concluded that the tidal effect of Earth would increase inversely with the cube of the distance of the moon, so that if the moon were at only half its present distance, Earth's tidal effect on it would be 2^3, or 8 times what it is now. If it were at one third its present distance, Earth's tidal effect would be 3^3 or 27 times stronger than it is now.

Roche decided that if the moon were at a distance of only 2.44 times the length of the Earth's radius, its *Roche limit*, the tidal effect would be strong enough to pull the moon to pieces. Since the Earth's radius is 6,350 kilometers (3,950 miles), the moon would have to be at a distance of 15,500 kilometers (9,610 miles) from Earth's center, which is only about one twenty-fifth of its present distance, to be broken up. (Of course, if the moon were that close, it would have a powerful tidal effect on the Earth, too, but since Earth has a stronger gravitational pull than the moon does, it would hold together under the strain.) If pieces of matter were to exist in the neighborhood of the Earth closer than its Roche limit, Earth's tidal effect would prevent them from coalescing into a large satellite such as the moon.

The Roche limit for Saturn is 2.44 times its radius, or 146,400 kilometers (91,000 miles). The rings of Saturn are entirely within this limit, so the material making them up could never come together to form a single sizable satellite. The smaller an object, the smaller the tidal effect upon it, so that small satellites are not broken up, and some can be found within the Roche limits of the outer planets.

For years, astronomers wondered why only Saturn had rings. Why not the other gas giants as well? In 1977, it was discovered that Uranus had rings, too. When Uranus passed in front of a star that year, the star's light was dimmed several times before Uranus actually reached it; there were rings of material that obscured it. However, they were so thin, sparse, and dark and reflected so little light that they weren't visible from Earth. When rocket probes went out to the giant planets and took photographs, the thin rings of Uranus were clearly seen. A thin ring was also discovered around Jupiter, and Neptune was found to have several.

Apparently, the giant planets all have rings, but why are Saturn's so much broader and brighter than the others? Does it have something to do with Saturn's strangely low density? Astronomers still don't know.

57. IS THERE LIFE ON VENUS?

In the last several decades, we have learned a great deal about the planets that we never knew and couldn't have known before, due in large part to new techniques involving radio waves (which I will discuss in greater detail later in the book) and rocket probes.

In 1974 and 1975, for instance, a rocket probe, *Mariner 10*, traveled past Mercury three times, taking photographs on each occasion. On the third pass, it approached within 327 kilometers (203 miles) of Mercury's surface. The photographs of the planet revealed a landscape very much like the moon's, with craters everywhere. Only three eighths of Mercury's surface was photographed, and in that region the largest crater was about 200 kilometers (124 miles) in diameter.

It was previously thought that Mercury rotated on its axis in eighty-eight days, the same period as its revolution about the sun, so that only one side faced the sun. It turns out, however, that it rotates in fifty-nine days, making just three rotations for every two revolutions.

It seems quite clear from Mercury's utter lack of air and water, and from its great heat (because it is only two fifths as far from the sun, on the average, as the Earth is) that there can be no life of our kind on it, or life of any kind, most likely.

But what about Venus? That seems to be another case altogether. Venus circles the sun in an orbit that lies between ours and that of Mercury. Its distance from the sun is not quite three quarters that of Earth, so we might expect it to be warmer than Earth, but perhaps not enormously so.

In 1761, the Russian scientist Mikhail Vasilyevich Lomonosov (1711–1765) was the first to note that Venus has an atmosphere. What's more, the atmosphere is filled with a thick, perpetual cloud layer that reflects about three fifths of the sunlight that falls on it—twice as much as Earth reflects. That would cool the planet somewhat, it seemed, so that it might well be suitable for life, especially since the clouds seemed to imply the presence of water on Venus, and perhaps large oceans of it.

Laplace's nebular hypothesis made it seem that Venus was given off by the condensing sun later than Earth was, so that it was a younger world. Science fiction writers frequently wrote of it, then, as a place where life was at an earlier stage than it was on Earth, a tropical paradise swarming with life, where dinosaurs were still the ruling animals.

After 1860, scientists learned how to analyze the light received from shining objects in order to work out the nature of the chemicals they contained (a process I'll discuss later on). Using such techniques, the American astronomer Walter Sydney Adams (1876–1956) detected carbon dioxide in Venus' atmosphere. As it happens, carbon dioxide is easier to detect than oxygen and nitrogen (the major constituents of Earth's atmosphere), so it might not be surprising that it was the first substance discovered there. Nevertheless, in our own atmosphere carbon dioxide makes up only about 0.03 percent of the whole, which would not be enough to detect easily, and it was natural to conclude that Venus might have considerably more carbon dioxide in its atmosphere than we have in ours.

The importance of this discovery lies in the fact that carbon dioxide absorbs infrared light much more than oxygen and nitrogen do. (Infrared light lies beyond the red end of the spectrum, and we cannot see it with our eyes, though we can detect it with our instruments.) A planet like Venus or Earth gets heat from the sun's visible light, which passes through oxygen, nitrogen, and carbon dioxide with equal ease. The planet loses heat at night in the form of infrared light, which passes through oxygen and nitrogen, but is absorbed by carbon dioxide. The infrared light heats the atmosphere slightly when it is absorbed and makes the planet warmer than it would be if it had no carbon dioxide. On Earth, this *greenhouse effect,* as it is called, is relatively small because there is so little carbon dioxide present. The warming is just enough to lift Earth out of an ice age and make the planet livable. On Venus, the additional carbon dioxide should create a greater warming effect, which would make it even warmer than was first suspected.

Every object gives off radio waves, which lie even farther beyond the red end of the spectrum than infrared does, and after World War II astronomers learned the techniques for receiving and analyzing the radio waves given off by objects in space. In 1956, a team of American astronomers, headed by Cornell H. Mayer, managed to receive radio waves emitted by Venus' dark side. The hotter an object, the more radio waves it gives off and the more energetic they are, and Mayer was astonished at both the quantity and energy of the radio waves he received. They seemed to indicate that the temperature of even the night side of Venus was far above the boiling point of water.

In 1962, the *Mariner 2* probe skimmed by Venus and measured the radio wave radiation  from it very accurately. Since then, other probes have done the same, and some have even landed on Venus' surface. The surface temperature on all parts of Venus is about 427°C (800°F), which is mainly because Venus' atmosphere is about 90 times denser than Earth's and is 98.6 percent carbon dioxide (Venus has 7,600 times more carbon dioxide in its atmosphere than Earth). These conditions have produced a runaway greenhouse effect.

At such high temperature, Venus is bone-dry. There is some water vapor in its clouds, but those also contain sulfuric acid. Venus is a totally unpleasant world, and there is no possibility of our kind of life on it. Nor does it seem possible that human beings will ever be able to land on it; all exploration will have to be done by nonhuman devices.

Radio waves can penetrate the cloud layer, however, and have enabled us to map the planet's solid surface and measure the speed of its rotation. These results led to another surprising discovery. It turned out that Venus was rotating on its axis very slowly—it takes 243 Earth days for it to make one rotational turn—and that is in the "wrong" direction, turning from east to west instead of west to east as the other planets do. We don't know why this is true.

In any case, we can cross off Venus as a possible abode for life.

58. IS THERE LIFE ON MARS?

Mars has always been the planet that most people would expect to find life on. It is about 50 percent farther from the sun than Earth is, so it is probably a cooler world, but perhaps not much cooler.

Mars has an atmosphere, but it doesn't have a perpetual cloud layer as Venus does, or even as many clouds as Earth does, so we can make out markings on Mars' surface. In 1659, Huygens followed these markings and showed that Mars, though distinctly smaller than Earth, rotated about its axis in 24½ hours, which is quite close to Earth's own rotation period.

In 1784, Herschel showed that Mars' axis is tipped toward the sun just about as much as Earth's, so that its seasons are probably similar to ours, except that each season has to be cooler than the corresponding season on Earth and nearly twice as long, since Mars is farther from the sun and takes 687 Earth days to complete one orbit around the sun. Herschel also spotted ice caps at Mars' north and south poles, which seemed to indicate the presence of water.

Early astronomers tried to map the markings on Mars, but they were not very successful at it, for no two could produce the same map. However, Mars approaches Earth more closely at some times than others, and every thirty years or so it makes its closest encounter, coming not much more than 56 million kilometers (35 million miles) away. Only Venus approaches us more closely, sometimes coming as close as 42 million kilometers (26 million miles). During Mars' closest approaches it can be seen most clearly, and of course, at each successive close approach, astronomers' instruments improved.

In 1877, such a close approach occurred, and an Italian astronomer, Giovanni Virginio Schiaparelli (1835–1910), produced the best map of Mars' markings that had yet been seen, and the first that other astronomers could agree with. Schiaparelli noticed that many of the dark markings on Mars were long and narrow. Astronomers before his time had also observed these markings, but Schiaparelli saw more of them than anyone else. Since the dark markings seemed to represent bodies of waters, Schiaparelli called them "channels." He used the Italian word for it, *canali*, and this was translated by British and American astronomers into the English word *canals*. This was an important mistranslation: A *channel* is a natural body of water, while a *canal* is man-made. As soon as astronomers began to talk of Martian canals, people began to imagine intelligent Martians building them.

It seemed to make sense. Mars, with its low surface gravity (only two fifths that of Earth), couldn't hold on to water vapor very well, and so it leaked away into space, making Mars an ever drying desert. To keep themselves alive and to keep their agriculture going, it could therefore be argued that a Martian civilization had built an intricate series of canals to bring water down from the ice caps to the warmer equatorial regions. It was a very dramatic picture that appealed to the general population—and to some astronomers, too.

The most influential supporter of the idea of Martian canals and life on Mars was the American astronomer Percival Lowell (1855–1916). He was a rich man who established a private observatory in Arizona, where the mile-high dry desert air and the remoteness from city lights made visibility excellent. From the observatory, he took thousands of photographs of Mars and made detailed maps that eventually included over five hundred canals. In 1894, he published a book entitled *Mars,* which advanced the fallacy that the planet harbored intelligent life.

A British writer, Herbert George Wells (1866–1946), used Lowell's book to write a novel called *The War of the Worlds,* which first appeared in 1898. In it he describes a Martian raiding expedition that comes to Earth for our plentiful water supply, with the intention of abandoning their drying planet and colonizing Earth instead. With the Martians' advanced technology, there was no hope that the Earth-people could stop the invasion, but in the end the Martians were defeated because their bodies could not fight off the activity of earthly bacteria. The novel was the first important picture of interplanetary warfare and was so well and frighteningly written that it convinced more people of the existence of life on Mars than Lowell's book did.

Not everyone accepted the notion of Martian canals, however. One American astronomer, Edward Emerson Barnard (1857–1923), who was noted for his excellent eyesight, never saw canals on Mars and insisted they were an optical illusion. Eyes, seeing small, irregular splotches of darkness, interpreted them as long, straight lines.

A British astronomer, Edward Walter Maunder (1851–1928), tested this idea. He set up circles within which he put smudgy irregular spots and placed schoolchildren at distances from which they could just barely see what was inside the circles. He asked them to draw what they saw, and they drew straight lines such as those Schiaparelli and Lowell had drawn on their maps of Mars.

Other astronomers also voiced their objections, but Lowell clung to his notions, and the general public went along with the drama of it. For more than fifty years after Wells's novel, science fiction writers seemed to be obsessed with Martian canals and intelligent Martians.

Gradually, though, scientific findings turned against the possibility of life on Mars. In 1926, two American astronomers, William Weber Coblentz (1873–1962) and Carl Otto Lampland (1873–1951), were able to measure the tiny quantities of heat that emerged from Mars and discovered that while the Martian equator might be mild in the sunlight, the Martian night was as cold as Antarctica. Such a large drop in temperature during a twelve-hour night gave the impression that the Martian atmosphere must be very thin.

In 1947, the Dutch-American astronomer Gerard Peter Kuiper (1905–1973) detected carbon dioxide in the Martian atmosphere, but could not find any oxygen or nitrogen. Not only might the Martian atmosphere be too thin to breathe, but its composition was such that it would be unbreathable even if it were denser. Hopes for intelligent life on Mars grew slim, indeed.

What was needed was a closer look, of course, and the coming of the age of rockets made that possible. In 1965, the *Mariner 4* probe passed Mars within a distance of 10,000 kilometers (6,200 miles) of its surface and took twenty photographs, which were sent back to Earth. No canals were visible in those photographs, only craters like those on the moon. What's more, *Mariner 4* sent radio waves through the Martian atmosphere, which turned out to be only a little over one two hundredth as dense as Earth's, with carbon dioxide its chief component.

The chance of intelligent life on Mars grew even smaller as other rocket probes took better and more detailed photographs. In late 1971, *Mariner 9* was placed in orbit about Mars and mapped the entire surface, revealing the existence of large but dead volcanoes, a huge canyon, markings on the surface that looked as though they might once have been riverbeds, and layered ice caps that might contain frozen carbon dioxide as well as frozen water. The temperature everywhere was far below freezing, and there were no canals; what had been seen were optical illusions just as Barnard and Maunder had maintained. Lowell had been completely wrong.

In 1976, two rocket probes, *Viking 1* and *Viking 2,* actually landed on the surface of Mars and took photographs showing an absolutely bleak and lifeless landscape. Automatic tests were made of the soil to see if there might be microscopic life present, but none was definitely found. We still can't say for certain there is no life on Mars or that there might not once have been, but there certainly seems to be no life on Mars today beyond the barest possibility of something equivalent to bacteria.

59. IS THERE LIFE IN THE OUTER SOLAR SYSTEM?

If Mars is too cold (among other things) for life as we know it, then surely the worlds that lie beyond Mars are even colder and less suitable. As for the four giant planets, the conditions on those worlds are so radically different from anything on Earth that we can't seriously expect to find our sort of life on them.

If we leave the giants out of consideration, then there remain various satellites, almost all of which are airless and almost all of which, if they possess water, do so only in the form of ice. They can all be eliminated out of hand, with two possible, though unlikely, exceptions, Europa and Titan.

The four large satellites of Jupiter—Io, Europa, Ganymede, and Callisto (in order of increasing distance from the sun)—are all subject to the large and forceful tidal influence of the giant planet. The satellites do not travel in perfect circles around Jupiter because of their pull on each other, and so as their distance from

Jupiter changes, they tend to stretch and contract slightly, a process that has a heating effect upon them.

Since the tidal effect increases inversely with the cube of the distance, as suggested by the findings of Edouard Roche, who first explained Saturn's rings, this pull is not very strong on the two outermost satellites, Ganymede and Callisto. These two have remained cold enough to retain icy material, so that they are larger than the other two satellites, and since Ganymede has a density of 1.9 and Callisto one of 1.6, they are probably mostly ice.

Io, the satellite closest to Jupiter, is subject to the strongest heating effect, so strong that it has no ice left and is an entirely rocky world, for its density is 3.6 grams per cubic centimeter. As a matter of fact, Io is heated to the point that the interior can break out in volcanic action. When the *Voyager 1* probe passed near Io in March 1979, eight volcanoes were seen actually erupting on its surface, and when *Voyager 2* passed in July 1979, six of the volcanoes were still erupting.

For the most part, what comes out of the Ionian volcanoes seems to be sulfur. This turns the surface of the satellite a red and orange color, while gusts of sulfur dioxide leave white patches. Any craters that formed in the bombardment of Io in the early days of the solar system are covered by the sulfur, so that the world is fairly smooth and lacks the craters that so mark up Ganymede and Callisto.

Europa, the second of the large satellites from Jupiter, is the smallest of the four, with a diameter of 3,138 kilometers (1,950 miles), somewhat less than that of our moon. Rocket probes show it to have a smooth surface, the smoothest of any world in the solar system. It is as though it were covered with a worldwide glacier.

But if the glacier was solid, its surface would be littered with craters, as Ganymede and Callisto are. Instead, it is crisscrossed by a large number of fine cracks, rather resembling the maps of Martian canals drawn by Lowell. The best explanation for them seems to be that occasional meteorites strike the glacier, which is merely an outer shell, and crack it, falling into an ocean of liquid below. (The liquid would be kept from freezing by the warming effect of Jupiter's tides.) The liquid water then wells up through the cracks formed by the meteorite and freezes there, leaving the surface unbroken.

The liquid might be mostly or entirely water, but even so, it does not have oxygen, and under the glacial covering there is no sunlight. Almost all of life on Earth depends on sunlight and oxygen. But only *almost* all. There are some forms of primitive bacteria that obtain their energy by bringing about chemical changes in sulfur and iron compounds that involve neither sunlight nor oxygen. In recent years, areas on the ocean bottom have been found where hot water wells up, rich in minerals that certain bacteria can use. Other, higher forms of life feed on the bacteria and on each other, and they seem to get along well. Is it possible, then, that Europa does have an ocean and that it can support a form of life there? Someday our instruments will have to be sent under the ice cover to check.

Some of the satellites in our solar system are cold and large enough to retain an atmosphere. (Cold gases are more sluggish in their molecular motions and more easily held by a weak gravity than warm gases are.) Thus, Triton, the large satellite of Neptune, was visited by *Voyager 2* in 1989, and it turned out to be a bit smaller than had been thought—only 2,730 kilometers (1,700 miles) in diameter and the smallest of the seven known large satellites. Even so, it was so cold ($-223°C$, or $-369°F$) that it does retain a thin atmosphere.

Triton's atmosphere is mostly nitrogen and methane, both of which freeze at very low temperatures, and as a result, its surface is slick with the ice of these substances. Still, there is enough heat on Triton to turn solid nitrogen into gas, so that every once in a while frozen nitrogen erupts in vapor form, shoving solid, icy material upward. Such *ice volcanoes* produce craters and ridges. Triton is the only world, besides Earth and Io, that seems to have live volcanoes, but there seems no reasonable chance for life there.

Pluto, which is considerably smaller than Triton, and Charon, its satellite, which is smaller still, also have thin atmospheres, but there's no reasonable chance for life there, either.

The satellite with the densest atmosphere is Titan, the largest satellite of Saturn. It is almost as large as Ganymede, with a diameter of 5,150 kilometers (3,200 miles), and its atmosphere seems to be even thicker than that of Earth.

As in the case of Triton, Titan's atmosphere is made up of nitrogen and methane. The methane is present in enough quantity and is close enough to the sun to be affected by sunlight. The more energetic radiation of the sun binds the molecules of methane (each of which is made up of one carbon atom and four hydrogen atoms) into more complicated molecules with several atoms of carbon each.

While methane is a gas at the temperature on Triton, the more complicated carbon compounds derived from it are liquid. It might be, then, that Triton has free liquid (a kind of gasoline, actually) on its surface. Triton's atmosphere is, unfortunately, so smoggy that its surface cannot be seen, but recently radio waves have been bounced off Triton's surface that seem to indicate liquid oceans with dry-land continents emerging from them. This is much like Earth, except that the oceans are composed of gasoline and are much colder.

Can life forms exist in gasoline? Again, we will have to send instruments to Titan's surface someday to find out.

The conclusion, then, is that except for Europa and Titan, each of which is an extreme long shot, there is no possibility of life in the solar system outside Earth. That's all the more reason why we should labor to save our lovely and unique world.

60. WHAT DOES THE SUN LOOK LIKE?

It is time that we turned to the sun itself, the center and the life-giving ruler of the solar system. But to ask what it looks like may seem too obvious a question. After all, doesn't everyone know what it looks like? It is a blazing circle of light.

Actually, it is too much a blazing circle of light, so bright that people cannot look directly at it for more than a second or so without damaging their eyes. As a result, it is very difficult to tell exactly what it looks like.

The sun's brilliance and its obvious importance as a source of light and heat have given it divine eminence in almost all mythologies. There are sun-gods everywhere. One of the best-known of these is the Greek sun-god Helios, though in the later myths it is Apollo who drives the blazing chariot across the sky every day.

The first monotheist we know by name was the Egyptian pharaoh Amenhotep IV, who came to the throne in 1379 B.C. and founded a new religion in which the sun (called *Aton*) was the only god. He changed his name to Akhenaton in honor of the sun, but the religion did not long survive his death.

Christianity does not, of course, give divine honor to the sun, but the sun was so identified as a symbol of the perfection of God that it, more than any other heavenly body, was considered perfect.

Occasionally, it *is* possible to look straight at the sun. Sometimes it shines through mist and can be looked at directly, and often at sunset it is dimmed sufficiently by the thick layers of dusty air its light must traverse that it can be gazed upon directly.

At such times, dark spots can occasionally be seen on its glowing surface. The Chinese astronomers noted these on numerous occasions and carefully recorded them. Undoubtedly, Europeans must have observed them, too, but they were never reported. The thought that the face of the sun could be marred by spots was too much an insult to the God that the sun symbolized, and it was easier to believe the spots were the result of some defect of vision.

Toward the end of 1610, Galileo and his telescope found there could be no mistake about it; there were definitely spots on the sun. What's more, they moved slowly and regularly across the face of the sun and showed that the sun was rotating on its axis once every twenty-seven days or so. Of course, this discovery created a great furor, and religious leaders were horrified at the possibility that the sun was desecrated with spots, but facts were facts, and Galileo won out (and made enemies in so doing).

Actually, the sunspots are not really black; they only seem dark compared with the glow of the sun. Every once in a while either Venus or Mercury come directly between the Earth and the sun and move slowly across its face (making what is called a *transit*). When this happens, the planets show up as extremely dark, black objects, and if they move near a sunspot, it is quite plain that the spots are dimmer than the sun itself, but are still bright.

In 1825, a German amateur astronomer, Samuel Heinrich Schwabe (1789–1875), began to study the sun and its sunspots. He spent seventeen years looking at it (with the proper precautions to avoid blindness) and discovered that the number of spots rose and fell in what seemed a cycle of ten years (more like eleven, according to continuing studies by others). This was the beginning of the science of *astrophysics*, the study of physical phenomena in stars and other objects in the universe. The reason for this *sunspot cycle* is not known even today.

This rise and fall of sunspot intensity seemed to have earthly importance, for in 1852, a British physicist, Edward Sabine (1788–1883), pointed out that the rise and fall in the intensity of Earth's magnetic field matched that of the sunspot cycle. This observation made it seem as if sunspots had something to do with magnetism, and in 1908, the American astronomer George Ellery Hale (1868–1938) found that there was a strong magnetic field associated with the sunspots. In fact, that made the sunspot cycle twenty-two years long, for during each successive eleven-year period the magnetic field reversed itself.

In 1893, Edward Maunder (one of those who doubted the tales of Martian canals) studied early records of sunspots and found, to his surprise, that between 1645 and 1715, there were virtually no reports of spots. Though he announced this finding, no one took it seriously, since early reports were thought to be unreliable anyway.

In the 1970s, however, the American astronomer Johan A. Eddy came across Maunder's report and checked into it. He considered not only the work done by people with telescopes in the early days, but went even further back to consider the naked-eye reports of the Chinese and others. He found that there are periodic *Maunder minima* and that the one Maunder reported was only the latest. What causes Maunder minima is still not known.

61. WHAT IS SUNLIGHT?

Most of what we know of the sun is the light we receive from it, so we should consider what that light can tell us. In the first place, sunlight seems to be simply white light, which seems to be the purest light, and it is fitting that we get this kind from the sun. But if so, however fitting it might be, it is unfortunate, for what information can we get out of something as pure and simple as white light?

Human-made light, on the other hand, is not necessarily white. The flames produced by burning wood and other fuels tend to be red, orange, and yellow and to lack the divinely pure quality, therefore, of the heavenly light of the sun. Furthermore, sunlight can be given color if it is made to pass through colored bits of glass, as in stained-glass windows. The results are very beautiful, but the color, it was felt, resulted from the addition of the impurities introduced to pure white light by man-made matter. Even the ruddy light of sunrise or sunset seemed to be the result of sunlight passing through dusty air. In fact, the only case of colored light that was visible seemingly without human or earthly intervention was the case of the rainbow, which was considered a divine product—a bridge used by the gods or a sign sent by God to promise there would never be another Flood.

In 1665, Isaac Newton investigated the nature of sunlight by allowing a ray of light to enter a darkened room through a chink in a curtain and passing it through a triangular piece of glass called a *prism*. The path of the light ray was bent as it passed through the prism, but it was not all bent the same way. Some parts of it were bent more than others, and the light that shone on the white wall beyond the prism was a rainbow. It was a band of color that started with red (the least bent part of the light), then orange, yellow, green, blue, and finally violet (the most bent part of the light), all the colors changing gradually and blending together. This was exactly the appearance and the sequence of colors of the rainbow.

Because the band of colors was an insubstantial phenomenon that had no mass, Newton called it a *light spectrum* (*spectrum* is from the Latin word for *ghost*). The rainbow, it would seem, is a natural spectrum formed when sunlight passes through tiny droplets of water that linger in the air after a rain.

Of course, some who might have argued that the colors were produced by the prism, even though the prism was itself colorless, but Newton took care of that objection by allowing the light ray that had passed through the prism and formed a spectrum to fall on another prism that had been turned in the opposite direction from the first. Now the light, instead of bending and separating as before, reversed itself and came together again. Out of the second prism came white light. It was clear, then, that sunlight was not pure, but a complex mixture of light of various colors. When those various colors act on the retina of our eye, the effect is what we call white light.

**62. WHAT ARE
SPECTRAL LINES?**

When Newton first studied the light spectrum, it seemed to him to be continuous. All the varieties of color in the spectrum changed from one to another without a break. But, in reality, the spectrum is not quite continuous, and there are some tiny gaps in which no color exists. Historians of science wonder sometimes how it was that Newton didn't notice this, but he was working with very crude instruments and the gaps might not have been very noticeable. In 1802, a British chemist, William Hyde Wollaston (1766–1828), *did* notice a few gaps in the spectrum and reported them, but he didn't consider them important and let the matter go.

Naturally, equipment for producing and studying a spectrum (the devices were called *spectroscopes*) improved. Eventually, the light was allowed to pass through a narrow slit so that the spectrum became a series of slits of different colors that all melted together, so to speak, to form a nearly continuous band. There were some colors missing, however, and where that color should have been, there was a dark slit, or dark line, crossing the bright spectrum.

In 1814, a German physicist, Joseph von Fraunhofer (1787–1826), was working with spectra produced by better equipment than anyone had yet used, and he found nearly six hundred such lines. (Modern physicists have located some ten thousand.) These were called *Fraunhofer lines* at first, but are now better known as simply *spectral lines.* As it turned out, these spectral lines proved to be of supreme importance.

Different chemical substances give off light of different colors when heated. Sodium compounds heat to give off yellow light; potassium compounds yield violet; strontium compounds, red; barium compounds, green; and so on. Such compounds are used in producing the spectacular fireworks with which all sorts of occasions are celebrated.

In 1857, a German chemist, Robert Wilhelm Bunsen (1811–1899), produced a gas burner so well-fed with air that it produced a virtually colorless flame. If it was used to heat a certain chemical, the light given off would produce a color not confused with any given off by the burner.

Bunsen's co-worker, the German physicist Gustav Robert Kirchhoff (1824–1877), used the *Bunsen burner* to produce light from different chemicals. He studied the spectra of such light and found that they were not continuous, but that each was made up of a few scattered individual colored lines. What's more, each different element (each different kind of atom) produced its own pattern of colored lines. The spectra thus provided a "fingerprint" for each element, and spectra could be used as a way of analyzing the elements present in a particular mineral.

When a sample of a mineral that has been strongly heated produces a series of colored lines that are not to be found in the case of any known element, the implication is that the mineral contains an unidentified substance. If the mineral was treated in various ways, fractions of it could be obtained in which the unknown lines were stronger, and the unknown element could be isolated, eventually, and studied. In this way, Kirchhoff discovered the elements *cesium* in 1860 and *rubidium* in 1861. They were named after the colors of the spectral lines that identified them, *cesium* from the Latin word for *sky blue,* and *rubidium* from the Latin word for *red.*

Kirchhoff went further in his study. When he shone sunlight through sodium vapor, the vapor absorbed certain portions of the sunlight and darkened some lines that were already there. He found that every vapor, if cooler than a light source, would absorb exactly those portions of the spectrum that it would emit if it were itself heated. In other words, you could identify elements (or simple compounds of those elements, for that matter) as bright lines against a dark background if the elements were heated and made to give off light, or as dark lines against a bright background if the elements were relatively cool and were absorbing light. It was through spectral lines, for instance, that carbon dioxide was first identified in the atmospheres of Venus and Mars.

63. WHAT IS THE MASS OF THE SUN?

We are now ready to consider what the sun is made of, but first, is the sun matter? The ancients considered it a mere ball of insubstantial light. It wasn't even earthly light. Aristotle thought that while the Earth was made of four "elements" (fundamentally different kinds of matter) including earth, water, air, and fire, the sun and the other heavenly bodies were made of *aether*, an unearthly substance characterized by the ability to glow forever. The very word *aether* comes from a Greek word meaning *to blaze*.

Even after it was understood that the sun was larger than the Earth, it might still be argued that it was insubstantial, unearthly, and without mass, so that its mere vast size was of no importance— the same point the early astronomers raised regarding the moon. But this uncertainty changed with Newton's law of universal gravitation in 1687, when it became clear that the Earth was bound to the sun by a powerful gravitational pull, and if the sun was the source of such a pull, it had to have mass.

But how much? This is not hard to determine. We know how long it takes the moon to turn around the Earth at a distance of 385,000 kilometers (239,000 miles). We also know how long it takes the Earth to turn around the sun at a distance of 150 million kilometers (93 million miles). From this, we can calculate how much more massive the sun is than the Earth. It turns out that the sun has a mass that is 330,000 times as great as that of the Earth. Therefore it is not an insubstantial ball of light, but a huge ball of matter about 1,038 times as massive as Jupiter, the largest planet. In fact, nearly 99.9 percent of all the mass of the solar system is in the sun.

Nevertheless, the sun is not as massive for its size as the Earth; its density is only about 1.4 grams per cubic centimeter, so that it is only one quarter as dense. Clearly, its chemical composition must be quite different from that of the Earth.

64. WHAT IS THE SUN MADE OF?

Well, then, just what is the chemical composition of the sun? That would seem to be an impossible question to answer. How could we possibly collect a sample of the sun for chemical analysis?

In 1835, the French philosopher Auguste Comte (1798–1857), searching for an example of a piece of knowledge forever inaccessible to human beings, pointed out that humanity could never possibly learn the chemical composition of the stars. He died at the age of fifty-nine. Had he lived four more years, he would have seen the very determination made that he thought could not be made—or at least the beginning of it.

The answer lay in Kirchhoff's discovery that elements gave off a spectrum of characteristic bright lines when heated, or a corresponding spectrum of dark lines when absorbing light. Thus, the hot surface of the sun sends out all kinds of light and would produce a continuous spectrum if the light reached the Earth untouched. Sunlight, however, passes through the sun's lower atmosphere, which is hot but not quite as hot as the surface. The atmosphere absorbs some of the light and produces the dark lines that Fraunhofer had discovered. From the position of those dark lines, the nature of the elements present in the solar atmosphere could be determined.

The Swedish physicist Anders Jöns Ångström (1814–1874) was the first to investigate this matter. In 1862, Angstrom pointed out that some of the dark lines in the solar spectrum exactly matched, in position, the dark lines that would be produced if light passed through hydrogen. The conclusion was that hydrogen was present in the sun.

With that, other astronomers began to study the sun's spectrum to learn additional facts about its composition. What is known now is that about three quarters of the mass of the sun is hydrogen, the simplest of all the elements, and almost all of the rest is helium, the second simplest. Together, hydrogen and helium make up about 98 percent of the mass of the sun.

Apart from hydrogen and helium, out of every 10,000 atoms

in the sun, oxygen makes up 4,300, carbon 3,000, neon 950, nitrogen 630, magnesium 230, iron 52, and silicon 35. The remaining eighty or so elements are present in even tinier traces. These findings utterly disproved the Aristotelian notion that the composition of heavenly bodies is fundamentally different from that of the Earth. It is quite obvious now that everything in the universe, that we know of, is formed of the same atoms (and subatomic particles) that Earth is.

65. WHAT ARE THE PLANETARY BODIES MADE OF?

Now that we know the general chemical makeup of the sun and realize that the large majority of stars (and of the dust and gas between the stars) have about the same makeup, we actually have the chemical composition of the universe. That is, we do if we assume that the stars and gas clouds make up most of the universe. (This assumption may not be accurate, by the way, as we shall see later in the book.)

We can now divide the types of matter in the universe into four general classes:

Gases. The two simplest elements, hydrogen and helium, make up about 98 percent of the universe. They are gases that are made up of very light atoms that move about very quickly. The less massive an atom is and the higher its temperature, the more quickly it moves. The more quickly atoms move, the harder it is for gravitational pulls to hold on to them.

This means that a hot body cannot hold on to hydrogen and helium unless it is very massive and has an enormous gravitational pull. The sun is massive enough to hold on to the hydrogen, helium, and all the other elements in the original cloud of dust and gas out of which it was formed.

If an object is cold, at least on the surface, it can hold on to hydrogen and helium more easily than if it was hot, and it need not be as large and as gravitationally powerful as the sun to accomplish that purpose. The four giant planets, Jupiter, Saturn, Uranus, and Neptune, are largely hydrogen and helium, and are sometimes called *gas giants* as a result.

These facts account for the low densities, about 1.4 grams per cubic centimeter, that the sun and the gas giants have. The density would be lower still if the interior portions of these large objects had not been compressed by pressure. The unusually low density of Saturn remains rather surprising.

Ices. A second type of matter is the ices, present in the universe in far smaller amounts than hydrogen and helium. These are made up of molecules that contain the secondary elements oxygen, nitrogen, and carbon, in combination with the atom of the overwhelmingly present hydrogen. Oxygen, combined with hydrogen, produces the molecules of water; nitrogen with hydrogen, ammonia; and carbon with hydrogen, methane. Water freezes to solid form at 0°C (32°F). Ammonia freezes at a lower temperature than that, and methane at a still lower temperature. There are also combinations of carbon with oxygen (carbon dioxide and carbon monoxide), combinations of carbon with nitrogen (cyanogen), combinations of sulfur with either hydrogen (hydrogen sulfide) or oxygen (sulfur dioxide), which can all be included among the ices.

Molecules of ice cling together more closely than the gas molecules. Small bodies can retain ices, even though their gravities aren't strong enough to hold very much hydrogen and helium. (The helium is usually lost altogether, for it can combine with nothing else. Some of the hydrogen remains, because it is able to combine with the other elements to form ices.)

The gas giants may well have mixtures of ices, minor in quantity compared with hydrogen and helium, but smaller bodies, if cold, are made up mainly of ices. These include the comets, for instance, and some of the satellites. Thus, Ganymede, Callisto, Titan, and Triton, four of the seven large satellites, seem to be made up mostly of ices.

Rocks. The rocky substances, a third type of matter, are formed by the combination of silicon with oxygen, magnesium, and other elements. There are less of these than of the ices, but they hold together even more firmly and are not dependent on gravity. The smallest pieces of matter made of rock can be held together by chemical forces even if the gravity of the object is negligible. Rocks also have high melting points and can survive even when quite close to the sun.

Some of the icy bodies might have rocky cores that contribute in a minor way to their structures. This might be true of the large satellites, for instance, and even of some comets. Small, hot bodies, including Mercury and the moon, lack both gases and ices and have bare surfaces of rock. Bodies such as the the moon, Mars, and Io are almost entirely rocky, though Mars is cool enough to retain some carbon dioxide and Io is cool enough to retain some sulfur-containing ices. Europa is intermediate, with a considerable quantity of surface ices surrounding an equally considerable rocky core.

Metals. Finally, iron mixes with metals to form a class of substances that is the least common of the four. Since the metals are denser than the other three classes of compounds, they sink to the center of the planet. Many of the rocky objects in the solar system might have comparatively small metallic cores, but the only worlds to have large metallic cores are Earth, Venus, and Mercury.

As you see, then, all the objects in the solar system, however different they seem chemically, could well have originated out of the same cloud of dust and gas. The differences we now see are the results of differences in temperature and mass.

66. HOW HOT IS THE SUN?

Surprisingly enough, the ancients did not emphasize the heat of the sun very much. They were so aware of it as a source of light that they tended to ignore it as a possible source of heat. We read descriptions of the sun-god driving a brilliantly luminous chariot pulled by luminous horses, but the heat of the system is not described. Then, too, very early tales of interplanetary travel described visits to the sun as well as to the moon, and while the brightness of the sun was understood, its heat wasn't mentioned.

And yet we all know that it is warmer in the day, when the sun is in the sky, than at night, when it isn't; warmer in the summer, when the sun is high in the sky, than in the winter, when it is low; and warmer in the sunshine at any time than in the shade. The question, then, is not whether the sun is hot, but *how* hot it is. The mere fact that we can feel its warmth at a distance of 150 million kilometers (93 million miles) is evidence that it is both a large and a hot fire. Fortunately, we don't have to stick a thermometer into the sun to find out what its temperature is. It turns out that both the quantity and the quality of the light produced by the sun depend on its temperature.

In 1879, the Austrian physicist Josef Stefan (1835–1893) showed that the total radiation of any object increases in proportion to the fourth power of its absolute temperature. (Absolute temperature is the temperature above absolute zero—the lowest temperature possible—which is $-273°C$, or $-459°F$.) If the absolute temperature doubles, the total radiation increases 2^4, or 16 times; if the absolute temperature triples, the total radiation increases 3^4, or 81 times; and so on.

Then, in 1893, the German physicist Wilhelm Wien (1864–1928) showed that the light produced by any hot object produced a peak of radiation somewhere in the spectrum and that this peak moved from the red end of the spectrum to the violet end as the temperature grew higher. The intensity peak in sunlight was in the yellow portion of the spectrum, and its precise location gives the temperature of the sun's surface. Thus, we know that the temperature of the sun's surface is about 6,000°C (9,500°F).

But that is only the surface. In the case of the Earth and, we have every reason to suspect, other planetary bodies as well, temperature goes up with depth. It would seem that the sun would grow hotter, too, as we imagine ourselves sinking below its surface. Since the surface of the sun is as hot as Earth's center to begin with and since the sun is much more massive and its center is subjected to far greater pressures than Earth's is, we might well expect that the center of the sun is considerably hotter than even the 50,000°C (90,000°F) that is thought to exist at the center of Jupiter. But how much hotter?

This problem was investigated during the early 1920s by the British astronomer Arthur Stanley Eddington (1882–1944). He began by assuming that the sun was a vast and extremely hot ball of gas that would act more or less like the gases we can study on Earth. Under the pull of gravity, the material of the sun ought to be drawn inward. In fact, if it was merely a gas, it would collapse rapidly to a relatively small size under the pull of gravity. (As we shall see, there are some conditions under which the sun would, indeed, do this.) Since, at present, the sun doesn't collapse, but maintains a size far larger than gravity would demand, there must be some force that acts to expand the sun's substance and to resist the tendency to contract.

The only phenomenon that Eddington (or anyone else) could think of that would do the trick was heat. When temperature goes up, a gas expands in volume, a fact that is known from experiments on Earth. Therefore Eddington felt that the sun was in a state of balance, with internal heat acting to expand it and gravitational force acting to contract it. In this balance, the sun would stay the same size year after year, indefinitely.

Eddington knew what the inward pull of gravity amounted to, so it was necessary only to calculate the temperature that would supply an outward push to equal it. Rather to his surprise, he discovered that the center of the sun had to be at a temperature in the millions of degrees. The usual figure now given is 15,000,000°C (27,000,000°F).

67. WHAT IS THE SOLAR CORONA?

During a total eclipse of the sun, the black disk of the moon is surrounded by a pearly light, called the *corona* (meaning *crown*), which is sometimes marked by beautiful streamers. At first, astronomers were not certain whether the light issued from the sun or the moon, but soon it was determined that it was definitely from the sun.

The corona is actually the upper atmosphere of the sun, which is a millionth as bright as the body of the sun itself and is therefore not visible except when that body is obscured by the moon. The corona, then, produces a light that is half the brightness of the full moon and keeps the world from being totally dark during an eclipse.

In 1931, the French astronomer Bernard Ferdinand Lyot (1897–1952) invented the coronagraph, an optical device that made it possible to observe at least the inner, brighter parts of the corona even when the sun was shining. That was the final proof (though none, by that time, was needed) that the corona was part of the sun.

The spectrum of the corona shows lines that were not in any substance studied on Earth. During a solar eclipse in 1868, which was visible in India, the French astronomer Pierre J. C. Janssen (1824–1907) observed such strange lines and referred them to an English astronomer, Joseph Norman Lockyer (1836–1920), who was an expert on spectra. Lockyer decided they represented a hitherto unknown element, which he named *helium*, from the Greek word for *sun*. This suggestion was not taken seriously till 1895, when the Scottish chemist William Ramsay (1852–1916) discovered helium on Earth. Helium is the only element that was discovered in a heavenly body before it was discovered on Earth.

There were other strange spectral lines in the corona, but they did not represent unknown elements. Instead, it turned out that atoms contained varying numbers of smaller particles called *electrons* and, under great heat, some of those electrons were lost. Atoms that had lost one or more electrons produced somewhat different spectral lines than intact atoms did, and in 1942, the Swedish physicist Bengt Edlén (b. 1906) identified some of the spectral lines of the corona to be atoms of calcium, iron, and nickel that had lost some electrons. For this to take place, the temperature of the corona had to be high—about a million degrees. This is attested by the corona's emission of high-energy radiation called X rays. However, the high temperature merely means that the individual atoms or atom fragments of the corona are very high-energy. There are so few of these spread out over so vast a space that the total heat of the corona is not high.

The solar corona has no sharp outer boundary, but continues to spread out through the entire solar system, thinner and thinner, so thin that it has no noticeable effect on the movement of the planets. However, the heat and energy of the sun drives charged particles outward in all directions. The American physicist Eugene Newman Parker (b. 1927) predicted this in 1959, and the effect was actually detected by rocket probes thereafter, notably by *Mariner 2*, which reached Venus in 1962.

This outward propulsion of charged particles is known as the *solar wind*, which travels from 400 to 700 kilometers (250 to 450 miles) per second. It helps keep comet tails pointed away from the sun. Its charged particles also strike the planets where the atoms accumulate, and if the planet has a magnetic field, as Earth does, the charged particles are trapped along lines stretching from the north magnetic pole to the south magnetic pole.

These charged particles in the Earth's neighborhood were first detected by rockets sent out in 1958 by a team under the American physicist James Alfred Van Allen (b. 1914). They were called *Van Allen belts* at first, but now are referred to as the *magnetosphere*. It was thought at first that these belts would interfere with spaceflight, but they have quite obviously not done so.

The charged particles leak down into the Earth's atmosphere near the magnetic poles and interact with the molecules there to produce streamers of colored lights, the *aurora borealis*, or *northern lights*, in the Arctic and the *aurora australis*, or *southern lights*, in the Antarctic.

68. WHAT ARE SOLAR FLARES?

In 1859, the British astronomer Richard Christopher Carrington (1826–1875) noted a starlike point of light break out on the sun's surface. At first he thought it might be a meteor striking the sun's surface, but actually, it was the first observation of what is now called a *solar flare*.

In 1889, the American astronomer George Ellery Hale (1868–1938) invented a device that made it possible to photograph the light of the sun by means of a single spectral line. This device easily picked up explosions on the sun's surface, and showed that these flares were not meteoric collisions but explosions that are associated with sunspots. We don't know exactly what causes solar flares, nor can we predict them, but they are more energetic than the comparatively untroubled solar disk. Sunspots are cooler than the rest of the sun (which is why they look darker), but their concurrence with solar flares means that the sun, when it is at sunspot maximum, is an active sun and more energetic than at sunspot minimum.

Solar flares produce particularly energetic gusts of solar wind. If the flare takes place near the center of the solar disk and is facing us, the energetic charged particles will reach Earth within a day or so, and an unusually high quantity of them will penetrate the Earth's atmosphere near the magnetic poles. This produces a *magnetic storm*, making the auroras particularly brilliant and wide-spread, upsetting the activity of magnetic compasses and radio waves.

Such a gust in the solar wind, if it catches astronauts unprepared, could kill them with radiation sickness. So far no astronaut or cosmonaut has been damaged by flares, but they do remain a threat.

69. WHY DOESN'T THE SUN COOL OFF?

Considering how hot the sun is now known to be, considering its powerful magnetic field, we need not be surprised at the existence of such energetic phenomena as a super-hot corona, the solar wind, and solar flares. But why doesn't the sun cool down?

This is a legitimate and puzzling question; after all, the sun pours down vast quantities of light and heat upon the Earth, but our small planet intercepts only a tiny portion of all the light and heat the sun produces—about one-hundred-millionth. Other tiny portions are intercepted by the other planets, but virtually all of it simply escapes into outer space beyond the planets.

The sun has been giving off these vast amounts of energy for

4.6 billion years without stopping, and it is still doing it. In fact, it shows every sign of continuing to do so for billions of additional years to come, without cooling down. How is this possible?

Actually, this question did not bother people before the mid-1800s. At that time, the law of conservation of energy was not entirely understood. The general feeling among the ancients was that the sun was merely a globe of light that kept shining eternally, or until the gods decided to blow it out. To be sure, there were earthly sources of light that continued to shine only as long as they were fed fuel. But that was merely earthly light. Divine light was considered to be something different.

In 1854, however, the German physicist Helmholtz, having worked out the law of conservation of energy seven years earlier, felt that it ought to apply to the sun as well as to earthly phenomena. He became the first, therefore, to question where the energy of the sun came from.

It was obvious that it couldn't come from ordinary sources, for at the rate the sun poured energy out into space, if it was simply a vast mixture of coal and oxygen, it would have burned up completely in 1,500 years. Everyone knows that the sun has been burning for far longer than 1,500 years—even according to the Bible it would have been shining for some 6,000 years. Helmholtz therefore turned to the process whereby the Earth and the other planets had gained *their* heat.

The sun had also probably formed by the joining of smaller pieces. Many more smaller pieces had to coalesce to form the sun than any of the planets, and far more kinetic energy was turned into heat in the process, which would explain why the sun was so much hotter than any of the planets. It merely poured out the energy it had gained in the early stages of its formation.

Helmholtz didn't know exactly how old the sun was, but he estimated that its age was many millions of years, and it seemed to him that the original supply of kinetic energy would not have been enough to keep it going all that time. It had to be continuing to gain kinetic energy as fast as it was losing heat energy.

He therefore considered the possibility that meteorites were

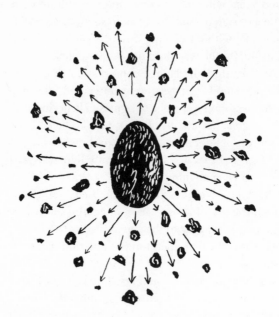

continually plunging into the sun as they were into the Earth. The sun was a far broader target than the Earth, with a far greater gravitational pull, which would subject it to many more meteorites.

That seemed a good notion, but it didn't work. As the meteorites plunged into the sun, they added mass to it, which would increase the sun's gravitational pull. It wouldn't increase by much, but the additional mass would be enough to make the Earth move just a little faster in its orbit and decrease the length of a year by a tiny but measurable amount. This gradual decrease in the year-length did not exist, however, so the meteorite theory had to be ruled out.

But then Helmholtz thought of something better. If the sun contracted as it formed from a vast cloud of dust and gas, why wouldn't it still be contracting? He calculated that even a very small contraction, one that was small enough to go unnoticed by the instruments of the time, would supply enough kinetic energy to keep the sun going. And it would not change the mass of the sun or the length of Earth's year.

If this were so, then the sun was slightly larger yesterday than it was today, and slightly larger still last year, and so on. Calculating backward, Helmholtz thought the sun had to be large enough to fill Earth's orbit 25 million years ago. That meant that Earth could not be more than 25 million years old.

This upset geologists and biologists, who had reasons for thinking that the Earth was much more than 25 million years old, but how could one argue against the law of conservation of energy?

Of course contraction was an insufficient explanation. But it wasn't till after the discovery of radioactivity, which took place two years after Helmholtz's death (and which was mentioned earlier in the book), that scientists came to realize that nuclear energy must be the source that kept the sun shining.

70. HOW DOES NUCLEAR ENERGY POWER THE SUN?

Deciding that nuclear energy powers the sun is easy, but figuring out just how the process works is not so easy. First, where does the nuclear energy come from?

In 1911, the English physicist Ernest Rutherford (1871–1937), considering the work he had been doing, which involved bombarding thin films of gold with beams of energetic radioactive radiations, found that most of the energetic radiations passed through the gold atoms as if there were nothing there, but a very small proportion of them bounced. From this he concluded that the atom was not just a little featureless ball, but had structure. At its center was an *atomic nucleus*, which was only one hundred-thousandth as wide as the atom itself. Virtually all the mass of the atom was in the nucleus, and around that center were one or more very light electrons, which I have mentioned before. The electrons made up most of the atom's volume, though they were pushed aside by the radioactive radiations as if they weren't there.

Ordinary chemical reactions (such as the burning of coal or oil or the explosion of TNT or nitroglycerine) are the result of shifts of the outer electrons from one atom to another. Such shifts tend to produce molecules with less energy content (like a ball rolling downhill—a ball possesses less energy in a low position than in a high one). When a chemical reaction takes place, the extra energy that is left over when the high-energy reactants become low-energy products appears as light or heat or explosive force.

The atomic nucleus is also made up of small particles. These are called *protons* and *neutrons*, and they, too, can rearrange themselves in ways that decrease their energy. The excess energy is then also liberated as radiation, heat, and so on.

These *nuclear reactions* occur much less frequently on Earth than chemical reactions and are far more difficult to start, stop, or alter in any way, so that until the end of the 1800s, they simply were not noticed. This was particularly true because the natural nuclear reactions that took place in connection with radioactivity were so slow that the amount of energy released in a given time was too small to be apparent.

The *total* amount of energy released in a nuclear reaction by a given quantity of material is enormously greater than the total amount released by the same quantity in a chemical reaction. Therefore, although chemical reactions, and even the kinetic energy caused by gradual contractions, are insufficient to keep the sun going for its lifetime, nuclear energy might do the trick, if scientists could only find the type of reactions involved.

The nuclear reactions that take place spontaneously on Earth involve the large atoms of uranium and thorium. Pieces of those atoms are chipped off in the course of radioactive breakdown, and energy is produced. Even more energy is produced if the uranium and thorium atoms are made to break more or less in half by a process known as *fission*. Yet even that amount of energy is not enough to power the sun, particularly since the sun contains only traces of these heavy atoms.

It is the atoms of intermediate size, however, that contain the least energy. In ordinary radioactivity, or in fission, the atoms slide "downhill," or give off energy, as the large atoms break up into smaller ones. The same would happen if very small atoms combined to form larger atoms. Suppose atoms of hydrogen (the smallest atoms) could be made to join together (undergo *fusion*) to form helium atoms, which are the second smallest. In that case, the energy produced by a given weight of hydrogen atoms undergoing fusion would be *far greater* than that produced by the same amount of uranium atoms undergoing fission.

Since it is now known that the sun is made up of three quarters hydrogen by weight and one quarter helium, it is tempting to suppose that the sun's energy arises from hydrogen fusion and that there is still plenty of hydrogen left to last it for billions of years.

There's a catch, though. The nuclei of heavy atoms are fairly unstable. It is as if they were on the edge of a cliff, so that it takes only the tiniest push, or even none at all, to send them sliding downward. Fission is, therefore, easy to start under the right conditions. On the other hand, hydrogen atoms have no natural tendency to fuse unless their nuclei come very close together, and this doesn't happen under ordinary conditions because there is an electron outside each hydrogen nucleus that acts something like a bumper. Two hydrogen atoms that collide bounce off each other's external electrons, and the two nuclei at the center of the atoms never come anywhere near each other.

That, however, is only the natural tendency under earthly conditions. At the center of the sun, temperatures are so high that hydrogen atoms are pulled apart and the hydrogen nuclei fly about on their own. The atmospheric pressure is so great that the hydrogen nuclei are pushed closely together, and since the high temperature means they are moving much more quickly than they ever move on Earth, they smash together with enormous force, which causes fusion to occur.

The German-American physicist Hans Albrecht Bethe

(b. 1906) worked on hydrogen fusion, studying the nuclear reactions that could be made to take place in the laboratory and calculating from these experiments what might happen at the temperatures and pressures in the sun's interior. By 1938, he had worked out a scheme for nuclear reactions that would suffice to power the sun, and in essence, his theory has been accepted ever since. Thus the question that Helmholtz first asked nearly a century earlier was answered.

71. ARE THERE STARS THE ANCIENTS DIDN'T KNOW?

Having discussed the planets and the sun, it is time to turn to the world of the stars. The question I begin by asking would have seemed silly in ancient and medieval times when the thought of invisible stars seemed a contradiction in terms. Stars shone and gave off light and therefore should be seen. Moreover, religious leaders of the Western world firmly believed that the universe was created entirely for the benefit of human beings. The stars were useful in astrological calculations concerning the future, and failing that, they were beautiful to contemplate. Invisible stars would be neither useful nor beautiful, would serve no purpose, and therefore could not exist.

And yet stars come in a wide variety of intensities. The brightest stars are so bright that no one who is not blind could fail to see them. The dimmest stars, however, are only about one hundredth as bright as the bright ones and can be seen only by people with sharp eyes. Isn't it possible, then, that some stars might be so dim that even people with the sharpest eyesight can't see them? If we think a little, there seems no reason why this shouldn't be so. Why should the dimness of stars stop just at the point where sharp vision can still make them out?

For the most part, people simply didn't reason out this problem. They were so attached to the idea that stars must serve humanity that the possibility of invisible stars was dismissed or, in fact, simply not imagined.

The coming of the telescope changed that. A telescope lens (or curved mirror) is much larger than the pupil of the eye, and it can gather light over a much larger area and concentrate it all at a focus. That means that stars look much brighter through a telescope than they would look to the eye alone, and if there was a star so dim that the eye could not make it out, a telescope might gather enough of its light to make it visible.

When, in 1609, Galileo turned his telescope on the sky, he found this to be exactly so. Wherever he looked he found many more stars than he could make out by eye alone. The sky, it seemed, was filled with myriads of stars that were too dim for the human eye to see unaided, but were there just the same, and visible through a telescope. That meant the universe was filled not with 6,000 stars, but with millions of them.

This simple feat of Galileo's accomplished two things. First, it was one more discovery that emphasized the size and complexity of the universe and showed it to be far from the simple structure it was assumed to be. Second, it was the first scientific discovery to make it perfectly plain that the universe did not necessarily exist only for the use and pleasure of humanity. Here were myriads of stars that, it seemed, could have no effect on human beings at all, and yet they existed. For the first time, it became possible for human beings to think of the universe as something indifferent to mankind, something that might have existed before man appeared on the scene and might continue long after he had left the stage. The universe was becoming much more magnificent, but somehow it was becoming colder and less friendly, too.

72. ARE THE FIXED STARS REALLY FIXED?

Here the answer might be, well, of course! How can anyone doubt that they are really fixed in place? After all, we look at the same stars in the same configurations that the ancient Sumerians looked at. There has been no change, and therefore the fixed stars are *fixed*.

Nevertheless, can we really say something is undergoing no changes just because we don't see any? Some changes take place so slowly that they might seem not to be taking place at all. Suppose, for instance, you stare at the hour hand of a clock for half a minute or so. You could easily come to the conclusion that it is simply not moving, that it is fixed in place. Yet if you leave and return an hour later, you find the hour hand *has* moved. It was pointing to *1* when you left, and to *2* when you returned.

Did it suddenly slip ahead when you weren't looking, or was it moving steadily, but too slowly to notice over a short period? If you decide to watch the hour hand patiently, not for half a minute, but for fifteen minutes, you will come to the conclusion that it is moving very slowly. If you look at it under a magnifying glass, you will see that even in half a minute it moves slightly.

Are we sure, now, that the fixed stars are truly fixed? Or are they moving so slowly (much more slowly than an hour hand) that their motion is undetectable unless we wait for centuries to pass? Even then, that might not be enough if we used our eyes alone. But the telescope (like the magnifying glass over the hour hand) can detect tiny shifts in position.

In 1718, Halley (who had worked out the orbit of Comet Halley), while checking the position of various stars with his telescope, found that three of them, Sirius, Procyon, and Arcturus, had changed position unmistakably since the ancient Greeks had reported their positions. To be sure, the ancient Greek astronomers did not have telescopes, but they were careful observers and they couldn't have been very far off.

In fact, those three stars had positions that were slightly different from those given by Tycho Brahe a century and a half earlier, and Tycho's observations were the best that could be had before the time of the telescope.

Halley could only come to the conclusion that these three stars had moved and changed their positions relative to the neighboring stars, and that they were continuing to move. This might possibly be true of all stars, so that the "fixed stars" were *not* fixed; they had a *proper motion*.

Nevertheless, the three stars that were detected as moving, although moving very slowly, had to have been moving more rapidly than other stars. What's more, those three stars were among the brightest in the sky. Was there a connection between the motion and the brightness? If so, astronomers might have to rethink the nature of the sky itself.

73. IS THERE A SPHERE OF THE STARS?

As I mentioned earlier, the ancients assumed that the sky was a thin, solid sphere enclosing the Earth, and that on it were the tiny, gleaming stars. All the discoveries until 1700 did not necessarily change that view. After Copernicus, it was no longer possible to suppose that the Earth was the center of the universe, about which everything revolved, but that only placed the sun at the center. The sky was still a celestial sphere that held the stars, but it was encircling the sun rather than the Earth.

Kepler's elliptical orbits had done away with the crystalline spheres of the planets, but the outermost celestial sphere of the stars still remained. Thanks to Cassini, the true scale of the solar system was discovered, and it turned out to be far larger than had been thought, but that only indicated that the celestial sphere was farther outward, too.

It was not until 1718, with Halley's discovery that the fixed stars were not fixed, that astronomers had to completely rethink their notions about the sky. It might be, of course, that the celestial sphere still existed even if the stars moved, and that the stars simply slid along the surface of the celestial sphere very, very slowly. But why should only a few stars travel quickly enough to notice after centuries had passed, and why did they happen to be the brightest stars?

It might be that some stars are larger than others, therefore brighter, and perhaps the larger stars are held to the sphere with somewhat greater difficulty, so that they slowly slip and slide along it. This, however, is merely an ad hoc reason, specifically invented to answer this puzzle, even though it doesn't fit into common experience and cannot be used to explain anything else.

On the other hand, some stars might be nearer to Earth than others. If so, the nearer stars would, on the average, seem brighter than stars that were farther away. Then, too, if the stars all moved at about the same speed, the nearer ones would seem to move more quickly—something, as I explained earlier in the book, that fits common experience. This explanation would make it clear why it happened to be bright stars that had detectable proper motion. The dimmer stars are moving also, but because they are so distant, they are moving so slowly in relation to us that a change in position cannot be detected in centuries, but perhaps only after many thousands of years.

If stars are at varying distances from the solar system, the celestial sphere cannot exist. Instead, space must be unlimited, with stars strewn through it like bees in a swarm. From 1718 on, the celestial sphere vanished from astronomical thinking, and the far grander picture of unlimited space took its place.

74. WHAT ARE STARS?

Originally, stars were thought to be what they looked like—little flecks of shining material attached to a solid sky, which was reasonable as long as the universe was thought to be relatively small and the sky not very far overhead. It became harder and harder to believe in stars as little flecks, however, as the universe began to seem larger and larger in the minds of astronomers.

By the time Halley discovered that the stars moved, it was clear that even the closest stars had to be billions of miles away if there was to be room inside the sphere of the stars for the huge solar system. For a speck of light to be visible at a distance of many billions of miles, how large would it have to be? If we think about it, we can't avoid coming to the conclusion that the stars must be *very* large objects.

The first person who had a glimpse of this fact, in 1440, was a German scholar, Nicholas of Cusa (1401–1464). It seemed to him that space must be infinite and that stars were strewn all through it. What's more, each star was an object equivalent to our sun, and each was associated with planets, on which life might exist. In all this, he was putting forward a startlingly modern view, but it was pure speculation and he had no hard evidence for it.

Once Halley had found that the stars moved, Nicholas of Cusa's ideas seemed inevitable. Halley wondered if Sirius, the brightest of the stars in the sky and therefore the closest or one of the closest, might not actually be as luminous as the sun. Perhaps it shone as no more than a point of light only because it was so far away.

How far away, then, would a sun like ours have to be in order for it to shine with no more than the light of Sirius? Halley made the calculation and decided that if Sirius were actually a sun as bright as our own, then its distance from us would have to be 19 *trillion* kilometers (nearly 12 trillion miles). A trillion, remember, is a thousand billion, or a million million—1,000,000,000,000.

Sirius, by Halley's computation, was 1,350 times as far from the sun as Saturn was. Stars that were dimmer than Sirius must by and large be even more distant. Again, the concept of the universe expanded, and it was no longer millions or even billions of kilometers across, but trillions.

75. HOW FAR ARE THE STARS, ACTUALLY?

Halley's estimate of the distance of Sirius depends on its being as luminous as the sun. This is a shaky assumption. It might, in actual fact, be dimmer than our sun, or for that matter, brighter. We need some more direct way of telling the distance of a star, so let's think about it.

The distance of Mars was estimated in 1672, with fairly good accuracy, by viewing the planet from Paris and French Guiana and calculating the parallax. Even the nearest stars, however, are pretty sure to be several hundred thousand times as far away as Mars, at the very least, which means that the parallax of the nearest stars would be several hundred thousand times smaller. Mars' parallax was hard enough to measure, even when viewed from different hemispheres; a star's parallax would be impossible to determine.

But there might be a way out of this dilemma. The Earth travels around the sun, and in six months it moves from one end of its orbit to the other, a distance of just about 300 million kilometers (186 million miles), about 23,500 times the width of the Earth. If a star is viewed from the same location first on January 1 and then on July 1, the parallax would be 23,500 times as great as it would be if it was viewed merely from opposite sides of the Earth.

Even under such conditions, the parallax of a star would be very small, considerably smaller than that of Mars as determined by Cassini. In fact, when Copernicus first advanced his theory, some astronomers pointed out that the stars showed no parallax and that therefore the Earth could not be changing position, but must remain in one place. Copernicus answered the objection, quite correctly, by saying that there was indeed a parallax, but the stars were so distant it was too small to measure. Without a telescope, it certainly was.

Still, if the stars were really located at vast and differing distances, their parallaxes could, in principle, be determined, and by the 1800s telescopes had at last improved to the point where the project was feasible.

In the 1830s, the German astronomer Friedrich Wilhelm Bessel (1784–1846) turned his telescope, the best that had ever been built, on a rather dim star called 61 Cygni. Even though it was dim, it had the largest proper motion known at that time, which led Bessel to assume, correctly, that it must be quite close, at least for a star. Finally in 1838, he obtained a tiny parallax and announced the distance of 61 Cygni. His first estimate was a little off, but it was excellent for a first attempt. The star 61 Cygni is 105 trillion kilometers (65 trillion miles) from Earth.

Very shortly thereafter, two other astronomers also observed the parallax of a star. This was not a coincidence, however, for as instruments improve, and sometimes even when attitudes change, a number of different scientists are often likely to make the same achievement at about the same time.

Two months after Bessel's announcement, the British astronomer Thomas Henderson (1798–1844) announced that the bright star Alpha Centauri was about 42 trillion kilometers (26 trillion miles) away. He had actually done his work before Bessel, but Bessel was the first to publish—that is, to make a written announcement—and it is the first to publish who gets the credit.

A bit later, the German-Russian astronomer Friedrich G. W. von Struve (1793–1864) showed that the bright star Vega was, to use the present-day figures, 255 trillion kilometers (158 trillion miles) away.

It turned out, by the way, that Alpha Centauri is the star that is nearest to us.

As for Sirius, that turned out to be about 82 trillion kilometers

(51 trillion miles) away—a little over four times as far away as Halley had estimated. The reason Halley had been wrong was that he had assumed Sirius to be as luminous as the sun, though it is actually 16 times more luminous.

These stars are all fairly close to the Earth. The vast majority are so much farther away that their parallaxes cannot be measured even by our best instruments today.

76. HOW FAST DOES LIGHT TRAVEL?

It is tedious to use large numbers; all those zeros are confusing. It is possible to deal in millions of kilometers or miles, or even a few billions, in dealing with the size of the solar system. But when we come to the stars and find we must deal with trillions of kilometers at the very least, and very likely with thousands of trillions, we can only throw up our hands and ask, What's the use?

The trouble is that kilometers and miles were designed to measure everyday distances on Earth, not vast astronomical distances. In order to work with the distances of stars easily, we need another kind of measuring unit, one that makes use of light.

To do this, we must ask, How fast does light travel? If you turn on the light in one corner of a room, how long does it take for the light to travel to the other end of the room and light that up, too?

The answer would seem to be, to anyone who has never thought about it, that light travels instantaneously, or with infinite speed. After all, when you turn on a light, every part of the room is instantly illuminated, and even if you turned on a powerful light in a vast stadium, the entire space would be lit up right away.

Nevertheless, *instantaneous* and *infinite* are difficult words, and it's possible that light doesn't spread out instantaneously, but merely in a very short period of time, one that is ordinarily too short to measure. Perhaps light does not travel at infinite speed, but only so quickly that it seems like infinite speed.

The best way to test this possibility is to try to make the light travel a very long distance. Then the time it takes to travel that long distance might be measurable. The first person who thought of this experiment was Galileo.

He and an assistant each carried a lantern on a dark night and climbed two neighboring hills. Galileo would open the little gate of his lantern and let a beam of light shoot out. The assistant would see it and instantly open the little gate of his own lantern and shoot a beam of light back at him. Galileo knew the distance between the hilltops, so the time that elapsed between the release of his own light and his sighting of his assistant's would be the time it took light to travel that distance twice—from one hilltop to the other and back.

There was indeed a small lapse of time. Part of it was due to the time it took light to travel, but part was due to reaction time. After all, it took some tiny amount of time for the assistant to realize he was seeing a flash and open the little gate of his own lantern.

Galileo therefore repeated the experiment with two hills that were farther apart. The reaction time would stay the same, so that any additional time between the first flash and the return would be entirely due to the time it took light to travel. He tried, and found there was *no* additional time; the time between flash and return was entirely reaction time. Light traveled too quickly to have its speed measured in this way.

Galileo realized it was necessary to find two hilltops that were much farther apart, but he knew that would be impractical. The bulge of the curved Earth would make one hilltop invisible from another one very far away. Besides which, Galileo couldn't find a flash bright enough to be seen for a very long distance. Of course, if he had some instrument that would measure extremely short intervals of time, he might be able to make a measurement, but he had no such device, so he gave up.

Then, nearly half a century later, the problem was solved quite accidentally. A Danish astronomer, Olaus Roemer (1644–1710), was studying Jupiter's four satellites. By that time the pendulum clock made it possible to measure time quite accurately, and it was known just how long it took each satellite to travel about Jupiter. At a certain time, in quite regular fashion, each one would disappear behind Jupiter and then reappear on the other side.

But it wasn't quite regular. During half the year, the satellite eclipses came a little ahead of schedule, and during the other half they were a little behind schedule. On the average, it all evened out but there were times when the eclipses were as much as eight minutes ahead of schedule, and half a year later, they were as much as eight minutes behind schedule.

Roemer tried to think of an explanation and realized that the eclipses were seen by sunlight reflected from Jupiter and its satellites, which traveled from Jupiter to Earth. As Jupiter and Earth orbited the sun, there were times when both planets were exactly on the same side of the sun, when light could travel from Jupiter to Earth along the shortest path possible. About two hundred days later, Jupiter and Earth were on opposite sides of the sun, and the light from Jupiter had to travel to where Earth would have been if it were on the same side and *then* across the full width of Earth's orbit to get to where Earth actually was.

It took sixteen minutes for light to travel across the width of Earth's orbit, eight minutes from Jupiter to the sun, then another eight to where Earth was on the other side. This distance was clearly much longer than the one between the two hilltops that Galileo had used. The two very distant "hilltops," Jupiter and Earth, were in sight of each other; the light was strong enough to be seen from one to the other; and the distance changed constantly with time. It was Galileo's experiment on an enormous scale, and this time it worked.

Roemer announced his result in 1676. He didn't have a completely accurate figure for the width of Earth's orbit, so his calculation was a bit low, but it was in the right ballpark. For the first time, people knew for sure that the speed of light was not infinite, but it was much faster than any other speed that had ever been measured. Other methods were devised to determine the speed of light more accurately, and the figure now accepted is just a trifle under 299,800 kilometers (186,300 miles) per second.

77. WHAT IS A LIGHT-YEAR?

How does the speed of light help us in talking about the distances of stars? Suppose we try to figure out how far light travels in one year. Each second it travels 299,800 kilometers (186,300 miles) per second, and there are 60 seconds to a minute, 60 minutes to an hour, 24 hours to a day, and 365¼ days to a year. That means there are almost 31,557,000 seconds in a year. Multiplying the distance light travels in one second by the number of seconds in a year, we find that in one year light will travel about 9.46 trillion kilometers (5.88 trillion miles). This distance is called a *light-year*.

The nearest star, Alpha Centauri, is 4.4 light-years away. This means that it would take light 4.4 years to travel from here to Alpha Centauri, or from Alpha Centauri to here. This gives you an idea of how far away stars are. It takes a beam of light ⅟₆₀th of a second to travel from New York to San Francisco, a little over

⅛ of a second to travel around the world, and about 16 minutes to cross Earth's orbit, but 4.4 *years* to reach even the nearest star.

Sirius is 8.6 light-years away; 61 Cygni is 11.2 light-years away; and Vega is 27 light-years away; and these are among the closest stars.

Though light-years are a very dramatic way of expressing long distances, astronomers don't use them much anymore. Instead, they measure distance in *parsecs.*

Every circle, including the enormous circle one can imagine drawn against the sky, is divided into 360 degrees, each degree into 60 minutes of arc, and each minute into 60 seconds of arc. That means that every circle is divided into 1,296,000 equal seconds of arc.

If you imagined a tiny *o* in the sky that was only 1 second of arc across and then imagined a whole series of such *o*s lined up so that they were touching in a line stretching across the sky, it would take 1,296,000 of them to form a complete circle around the sky. Each *o* is thus very small, indeed.

How far must a star be to have a parallax that would shift from its normal position first to one side, then the other, by 1 second of arc as the Earth moves around the sun? This answer is 3.26 light-years, a *parallax second,* or a *parsec,* for short. No star is that close, so every star we know has a parallax of less than one second of arc when seen from opposite sides of Earth's orbit, which is why it took so long to measure their distances. Alpha Centauri is 1.35 parsecs away, Sirius is 2.65 parsecs away, 61 Cygni is 3.44 parsecs away, and Vega is 8.3 parsecs away. One parsec is equal to just a little over 30 trillion kilometers (19 trillion miles).

78. IS THE SUN MOVING?

Since Copernicus' time, the sun came to be viewed as the unmoving center of the universe. Once Halley had discovered that the fixed stars moved, and once he began to speculate that the stars are really suns located at tremendous distances, it began to seem unlikely that our sun would be the only star that didn't move, and still more unlikely that objects uncounted trillions of miles away should be circling our sun as the center of all.

If all the stars are moving, why shouldn't the sun move, too? There is nothing unusual about it that we know of, except that it just happens to be much nearer to us than any other star. So we should assume that it is moving and ask, How can we *show* that it's moving, and in what direction?

In 1805, after more than twenty years of study, Herschel (who had discovered Uranus) felt he had the answer to the question. After all, suppose the sun were surrounded by stars in every direction, all separated, on the average, by equal distances. It would nevertheless seem that those nearest the sun were farther apart than those far away from the sun. (We see this effect if we're standing in a forest, where the trees near us are well separated, while those far away seem to be packed together very closely.)

Herschel measured the proper motions of as many stars as he could, and he found that in one particular direction, the stars seemed to be separating and moving away from a particular spot in the constellation Hercules. This spot Herschel called the *apex*. On exactly the other side of the sky, the stars seemed to be moving together toward a spot opposite the apex.

There is no reason the stars should be behaving in this peculiar fashion if the sun was standing still. But if the sun happened to be moving toward the apex, then the stars near the apex would be coming closer to us as the sun approached them and would seem to be spreading apart. The stars on the opposite side of the sky would be getting farther from us as the sun moved away from them and would therefore seem to be coming together.

The sun, Herschel concluded, was moving in the direction

of the constellation Hercules. After thousands of years of assuming the Earth to be the center of the universe, and then after two and a half centuries of assuming the sun to be the center of the universe, it turned out that, as far as astronomers could tell, there was *no* center of the universe. Everything was moving.

As a matter of fact, Nicholas of Cusa, among all the other correct guesses he made concerning the universe a century before Copernicus, had also maintained that there was no center to the universe.

79. ARE THE LAWS OF NATURE THE SAME EVERYWHERE?

In talking about the origin of the solar system, I dealt with such subjects as the law of gravitation, the law of conservation of angular momentum, and the centrifugal effect. I said that it was safe to suppose that these rules were legitimate because they worked on Earth here and now.

But how do we know that because something works now, it worked 4.6 billion years ago? How do we know that because something works here, it will work on other worlds? In short, how do we know that the laws of nature are the same throughout space and time?

Why should the laws of nature be different at different times or at different places? They certainly don't vary from place to place on Earth, and they haven't changed during the last few centuries when scientists have been investigating matters in detail.

This argument is not very convincing, though, for what are a few thousand kilometers and a few hundred years when we have to wonder about distances of many light-years and times of billions of years?

But if the laws of nature weren't universal, we would encounter many phenomena that we couldn't understand. There would be chaos and anarchy in the universe, because the rules we think we know wouldn't hold true under a variety of conditions.

However, maybe that's the way it is. There are, indeed, many phenomena throughout the universe we don't understand even today, and perhaps we are indeed faced with chaos and anarchy. In recent years, as a matter of fact, scientists have decided that some aspects of the universe are more chaotic than had been suspected.

Still, scientists generally like to think that the universe is essentially simple and that the same laws of nature hold true everywhere and throughout time, but this is just a comfortable assumption. Before we can believe it, we must have examples and evidence.

For instance, at the end of the 1700s, the most important generalization about the physical world that man had yet discovered was Newton's law of universal gravitation. There was no

doubt that it worked throughout the solar system, for all the planets and satellites moved almost precisely in accordance with it. When it turned out that Uranus' motion didn't conform to it exactly, astronomers suspected that another planet might exist beyond it, whose gravitational effect explained the discrepancy. Such a planet, Neptune, was searched for and found, just where it had been predicted.

As long as it was assumed that the solar system was virtually all the universe there was, universal laws were satisfactory, but once it turned out that the stars were suns located extremely far away, astronomers grew uneasy. Would the laws of nature hold at such unimaginable distances?

Herschel answered this question also. He was looking for evidence of the existence of parallax among stars, and it occurred to him to study stars that were very close together in the sky. At the time, it was taken for granted that all stars existed, like our sun, in lonely splendor. Therefore if two stars seemed close together in the sky, it was only because they lay in the same direction from us and one was much farther off than the other. In that case, the closer of the two might show a tiny parallax relative to the other.

He found that in the case of such stars there were tiny shifts in position, though not the kind of shifts one expected of parallax. By 1793, he was convinced that he was watching pairs of stars, *binary stars,* that were close together in reality and not merely in appearance and that were circling each other. Such stars were bound by gravitational pull, and from their motions it could be shown that Newton's law of gravitation, which had been deduced from the motion of the moon around the Earth, applied not only to all the bodies of the solar system, but to the distant stars as well.

This was the first indication that stars did not necessarily exist singly; they came in pairs, and as it eventually turned out, in more complex associations, too. Before Herschel's death, he had located no fewer than eight hundred binary stars. All without exception obeyed the law of gravitation as worked out by Newton and made more general by Einstein.

So it has gone. In the last two centuries, all scientific discoveries have fit the notion that the laws of nature apply everywhere throughout space and time. There might be special conditions of extreme nature where the laws break down, but we have not yet been able to study such conditions adequately. It might also be that, as scientists have recently come to think, there are chaotic conditions that cannot be predicted or explained with confidence, but such chaotic conditions rule everywhere alike, here on Earth and on the farthest star.

80. WHAT ARE VARIABLE STARS?

The Aristotelian belief that the objects in the sky were eternal and changeless seemed reasonable. The stars certainly looked absolutely the same night after night.

Yet this was not entirely true. Consider the case of the second brightest star in the constellation Perseus, Beta Persei. Every two days and twenty-one hours, it loses more than half its brightness, and then after a short interval regains it.

This might have been noticed in ancient and medieval times. The constellation Perseus shows that hero of Greek myth at the moment he has cut off the head of the snake-haired Medusa. He is lifting the severed head, which is marked by Beta Persei, so that the Arabs (and we today) call the star Algol, the Arabic word for *the ghoul.* Yet no one mentioned this light variability before modern times. It might well be that the changing brightness, a sign of the impermanence of objects in the sky, having been noticed, made people so uneasy that no one would speak of it.

In 1782, the English astronomer John Goodricke (1764–1786), a brilliant deaf-mute who died young, suggested that Algol was a binary star and that one of the pair was rather dim. Every two days and twenty-one hours, the dim star moved in front of the bright one and eclipsed it, thus accounting for the temporary loss in brightness. When the dim star moved out of the way, the brightness was restored. Goodricke was ahead of his time, for the actual existence of binary stars had not yet been demonstrated by Herschel. In time, however, he turned out to have been perfectly correct.

There are a number of such *eclipsing variables,* but there are also stars whose brightness changes with time in an irregular fashion. In 1596, the German astronomer David Fabricius (1564–1617) detected a star in the constellation Cetus, Omicron Ceti, which varied in brightness. As astronomers continued to watch it, they noted that it was sometimes bright enough to be one of the hundred brightest stars in the sky and sometimes grew so dim it could not be seen without a telescope. This dimming and brightening occurs in cycles of almost a year, though its magnitude is so irregular that the effect cannot be the result of an eclipse. The conclusion is, then, that the star simply radiates more light and heat at one time than another. It is a true *variable star*, and surprised astronomers named it Mira (Latin for *wonderful*).

In 1784, Goodricke discovered still another kind of variable star, Delta Cephei, in the constellation Cepheus. It has a regular variation in brightness, but this is not the result of an eclipse because the increase in brightness is quick and the decrease is slow. (If it were the result of an eclipse, the increase and decrease would take place in equal times, as in the case of Algol.)

Hundreds of other stars with this same pattern of rise and fall in brightness have been discovered, and they are lumped together as *Cepheid variables*. Some Cepheids complete their variation in three days, and some take as long as fifty days. As I will explain later, Cepheids turned out to be enormously important as a way of measuring huge distances.

81. HOW DO STARS DIFFER FROM ONE ANOTHER?

Until modern times, the chief way in which stars appeared to differ was their brightness. Hipparchus was the first to divide the stars into classes depending on their brightness. The twenty brightest stars of the sky are of the *first magnitude*. Then, in order of decreasing brightness, there are stars of the second, third, fourth, and fifth magnitude, while those of the sixth magnitude are just visible to the unaided eye.

The brightness of a star can be measured with such delicacy that magnitudes can be measured in decimals. A star can be of magnitude 2.3 or 3.6, with each degree of magnitude representing a brightness 2.512 times that of the next higher magnitude. A star of magnitude 2.0 is 2.512 times brighter than one of 3.0, and so on.

Some of the first-magnitude stars are so bright that they must be given numbers of 0 magnitude, or even go into negative numbers. Sirius, the brightest star in the sky, has a magnitude of −1.47. The scale of magnitudes can also be applied to objects besides stars. Venus, at its brightest, has a magnitude of −4; the full moon is −12; and the sun is −26. The magnitude range can be extended to dim stars that can only be seen through a telescope, so that some stars have magnitudes of 7, 8, and so on, all the way up to 20 and beyond.

One star might be brighter than another, not because it radiates more light, but because it might be closer to us. A relatively dim star that happens to be near to us might appear to be brighter than a star that is actually much brighter, but happens to be located much farther away.

If you know the distance of a star and its magnitude, then you can calculate the actual brightness, or *luminosity,* of that star. You can also suppose any star is at a standard distance of 10 parsecs (32.6 light-years) and calculate how bright it would look in the sky at that distance, a measurement known as the *absolute magnitude.*

For instance, if our sun were 10 parsecs away, it would have a magnitude of only 4.6, so it is not, in truth, a very luminous star. Sirius, at that same distance, would have a magnitude of 1.3, so it is considerably more luminous, and there are stars that are even more radiant. The star Rigel in the constellation Orion has an absolute magnitude of −6.2 and is about 20,000 times more luminous than the sun. The very luminous stars are rare, however. They are noticeable because they tend to be bright, but they are not numerous, and about nine tenths of all the stars are less luminous than the sun.

In 1914, the American astronomer Henry Norris Russell (1877–1957) showed that stars can be arranged in orderly progression—or that 95 percent of them can be. The greater the mass of a star, the more luminous and hotter it is. Most stars, arranged in order of mass from small, cool, and dim to large, white-hot, and bright, can be sorted according to a *main sequence.*

Eddington, who figured out the central temperature of the sun, explained the nature of the main sequence. The more massive a star, the more forcefully gravitation pulls its material inward, and the higher its central temperature must be to balance this force. The higher the central temperature, the more light and heat the star emits. In other words, the more massive a star, the more luminous it must be, a principle known as the *mass-luminosity* law.

A star's temperature rises faster than its mass, so that if a star is massive enough, the internal temperature is so great and the expansive push outward is so powerful that the star becomes unstable and is liable to explode. For this reason, stars with more than 60 times the mass of the sun are not likely to exist.

On the other hand, the less massive a star, the less temperature is required at its core to balance its modest gravity. If the star is small enough, the temperature at its center is so low that it does not shine at all. An object with less than a tenth the mass of the sun would be dark and would not be a star in the usual sense of the word.

Such failed stars might still be a hundred times the mass of Jupiter, however. They would be warm and would radiate infrared light, which is less energetic than visible light. They are called *brown dwarfs* and are difficult to detect, but astronomers are searching for them, for it is conceivable that they might exist in great numbers and might affect the nature of the universe, if so. As long as a star maintains a good hydrogen supply as part of its structure and continues to produce radiation by hydrogen fusion, it remains on the main sequence.

82. WHAT HAPPENS WHEN THE HYDROGEN SUPPLY OF A STAR RUNS LOW?

Once scientists had decided that stars, including our sun, produce their energy by hydrogen fusion, this became an important question. The sun and stars generally contain a vast quantity of hydrogen, but the supply is not infinite, and it will not last indefinitely. What happens, then, as the hydrogen supply dwindles?

It might seem that as the hydrogen supply begins to give out, a star will produce less and less energy. It would cool down and be unable to counter the pull of gravity, so that it would eventually shrink in size and become a cold, dense object—a dead star. Actually, this might happen in time, but a number of startling intermediate stages must occur before the star's ultimate demise. This theory of stellar classification first appeared in the work of the Danish astronomer Ejnar Hertzsprung (1873–1967), who first advanced the concept of absolute magnitude.

Hertzsprung found that some stars that delivered light of a reddish color had high absolute magnitudes and therefore were very dim. Others had low absolute magnitudes and were very luminous. He found nothing in between.

If a star delivers light of a reddish color, that is an unfailing sign that its surface is comparatively cool, with a temperature of not more than 2,000°C (3,600°F). Such a star, if it belongs to the main sequence, is bound to have a low mass, so it is called a *red dwarf*. Red dwarfs abound in the universe; three fourths of all stars seem to be of this type.

The puzzle was the bright red stars. The surfaces of such stars had to be cool, so that each unit portion of the surface delivered much less light than each unit portion of our sun, even though they were far more luminous. The only explanation for this seemed to be that although a given portion of the surface is dim, there is an enormous amount of surface. In other words, the bright red stars are much, much larger than the sun, which accounts for their high luminosity. These were called *red giants*.

At first it was thought that red giants were stars in the process of condensing, very young stars that would grow smaller and hotter and then continue to condense and grow dim until they had become red dwarfs. But this could not be, because they release too much light and heat to be merely condensing into stars. They had to already have nuclear furnaces in place at their cores. As astronomers continued to study the hydrogen fusion process at the center of stars, they found that red giants were not at an early stage in stellar evolution, but at a late one.

Astronomers found that as hydrogen fuses to helium, the helium collects at the center of the star to form a *helium core*. Then hydrogen fusion continues at the outer circumference of the helium core. This core becomes more massive and more compressed, and its temperature slowly increases, so that with time a star heats up rather than cools down.

Finally, the temperature at the core becomes so high that the helium begins to fuse into more massive atoms such as carbon and oxygen. Then the heat produced by the star through helium fusion, on top of the still ongoing hydrogen fusion, becomes greater than is required to balance the star's gravitational pull inward, and it begins to expand. As it expands, the outer layers cool as the heat produced is spread out over a larger and larger area. Each individual bit of surface area cools down, so that the star turns red, but the total heat dispersed over the entire swollen surface is higher than it was before the swelling began.

Some stars, as they expand, do so intermittently, expanding for a period, then contracting, then expanding, and so on, over and over again, though eventually expansion will prevail. These expansions and contractions are represented by the Cepheid variables. When a star expands to a red giant, it is said to have "left the main sequence."

The best-known red giant is the star Betelgeuse in the constellation Orion. It is estimated to have a diameter of 1,100 million kilometers (700 million miles), so that it is 800 times as wide as our sun. If Betelgeuse were shining in place of the sun, it would be so large that its swollen body would incorporate the entire inner solar system. Its surface would lie beyond Mars and into the asteroid belt.

83. WILL OUR SUN EVER BECOME A RED GIANT?

It will have to as its hydrogen supply begins to run low, but this does not represent an immediate danger. The sun should stay on the main sequence altogether for about 10 billion years. Since it is only 4.6 billion years old, it is merely middle-aged. Of course, it will gradually grow warmer, and during the last billion years or two that it spends on the main sequence, the Earth might be too hot for life. But that still leaves us about 3 billion years, and it is very doubtful that the human species will endure for even a small fraction of that time.

Of course, if we do survive and learn to adjust to the increasing temperatures, after about 5 billion years have passed, the sun will begin to expand. It is considerably less massive than Betelgeuse, so it won't expand as far, but it will grow large enough to destroy the Earth. Unless we, or our distant descendants, manage to transfer to a planetary system circling another star or learn to live in space independently of stars and planets, that will be the end for us.

Different stars remain on the main sequence for different lengths of time, depending on their mass. Remember, Eddington found that the more massive a star, the greater the quantity of heat it must produce to counter its greater gravitational pull, and the quantity of heat must rise faster than the mass does. This means that a giant star, with a large supply of hydrogen, must expend it so rapidly that it remains on the main sequence for a far shorter time than a dwarf star, which consumes its lesser supply of hydrogen in tiny quantities. In other words, the more massive a star, the shorter its time on the main sequence.

A star the mass of our sun might endure on the main sequence for 10 billion years, but a small red giant, just hot enough to be a reddish gleam, might last for 200 billion years on the sequence. Very luminous stars, on the other hand, are short-lived. The largest and most luminous are not likely to stay on the main sequence for more than a few million years.

84. WHY DO VERY LUMINOUS STARS STILL EXIST?

That is a good question. If giant stars are very short-lived, why do we still see a number of them on the main sequence? Why haven't they left the main sequence and expanded into red giants long ago? For instance, the star Sirius is about three times the mass of the sun and is using up its hydrogen about twenty times faster than the sun. Thus it should remain on the main sequence for only about half a billion years altogether. If Sirius had become a star at the same time as the sun did, 4.6 billion years ago, it would have become a red giant 4 billion years ago, yet in actual fact, it hasn't become a red giant even today.

The only reason we can possibly advance to account for this is that Sirius became a star less than half a billion years ago and in that brief time has not yet turned into a red giant. Similarly, the very brightest main-sequence stars we see in the sky today must have formed just a few million years ago or they would be red giants now.

This means that all of the stars did not form together when the universe as a whole came into being. Some small stars developed in the very early days of the universe and might still exist on the main sequence today, while others were produced in different sizes and have stayed on the main sequence for briefer periods, sometimes for very brief periods, and then left it, while still others have formed only recently.

We are quite certain that the sun itself is not as old as the universe. When our solar system was created, the universe had already existed and probably looked very much as it does today. (We'll discuss the question of how old the universe might be later.) In fact, there is no reason to suppose that stars are not in the process of formation right now.

The trouble is that it is very difficult to catch star formation in progress. First of all, stars form inside large clouds of dust and gas, and we can't easily penetrate those clouds to see exactly what is happening. Then the formation requires a period of time that might be very short in terms of astronomical existence, but is quite long compared with our lifetimes. If it takes a million years for parts of a cloud to collapse into a new star, then even during the entire course of astronomical investigation since the invention of the telescope, we don't see much happening. Nevertheless, astronomers are quite sure that new stars are being born right now.

85. WHAT IS A WHITE DWARF?

Once a red giant is formed, most of the available fusion energy that would allow it to maintain a quiet life is gone, especially since it is expending it now at a faster rate than ever. After a few million years at most, it can no longer keep itself expanded against the gravitational pull.

If we stop to think, we can see that this must be true, for if red giants remained red giants for long periods of time, they would litter the sky. Every massive star that had ever existed would eventually become a red giant and stay there. In actual fact, however, red giants are few in number, which means they must disappear (as red giants, anyway) after a relatively short period of existence.

When a red giant no longer has the energy required to remain expanded, it must collapse, though not to the size it had as an ordinary star on the main sequence, but further still, into a new and more extreme kind of dwarf star. Astronomers became aware of the existence of such dwarf stars long before they came to understand how stars changed with time (*stellar evolution*) and even before the discovery of red giants.

In 1844, F. W. Bessel, the first astronomer to announce the true distance of a star, was studying the motion of Sirius. Ordinarily, stars, in their proper motion, move very slowly in a straight line. But this was not the case with Sirius, which Bessel found moved in a wavy line. Bessel thought about this eccentricity and came to the conclusion that the only force known that could pull a star noticeably out of its line of travel was the gravitation of another star.

Suppose Sirius was not a single star, but a binary one. Sirius and its companion would be moving through space together, but as they did so, they would also be circling each other about a common center of gravity, and it would be this center of gravity that marked out a straight line through space. Sirius would be first on one side of the center of gravity, and the companion would be on the other; then they would change places. If Sirius and its companion circled their center of gravity every fifty years and if Sirius was about 2½ times as massive as the companion, that would account for Sirius's wave path.

But why couldn't Bessel see the companion? The logical conclusion was that the companion was a burned-out star. People had no idea, at that time, what the source of a star's energy might be, but whatever it was, Bessel thought it had been used up, and the companion, dark and cold, but with all its original mass, must be circling about the center of gravity. It came to be called the "dark companion," and Bessel later found that the star, Procyon, had a dark companion, too.

Then, in 1862, the American astronomer Alvan Graham

Clark (1832–1897), while testing a new telescope, noted a spark of dim light near Sirius. At first, he thought it was a flaw in the telescope, but further study showed that he was seeing a dim star. He was in fact seeing Sirius's dark companion, which had a magnitude of 7.1. This was not bright enough to see without a telescope, only perhaps ⅛,₀₀₀ as luminous as Sirius, but it was not cold and black. It came to be called Sirius's "dim companion." More properly, it is called Sirius B, while Sirius itself is Sirius A.

In 1896, the German-American astronomer John Martin Schaeberle (1853–1924) detected Procyon's companion. Now called Procyon B, it had only half the mass of Sirius B and was even less luminous.

Now let's think about Sirius B. From its effect on Sirius A, it was eventually determined that it must have a mass just about equal to that of our sun, but its luminosity was only about ⅟₁₃₀th as great.

A few decades later, when the relationship of mass and luminosity had been worked out, that would have been puzzling, indeed, for a star with the mass of the sun should have had the luminosity of the sun. In the early 1900s, however, this was not yet understood, so it didn't bother astronomers.

What did bother them was that if Sirius B was so much less luminous than the sun, it ought to be cooler, and shine with a red light. Instead, it shines with a white light that is just like that of Sirius A. What was needed was the spectrum of Sirius B; from the distribution of colors of light and from the dark lines present, the surface temperature could be determined.

In 1915, W. S. Adams, who was the first to detect carbon dioxide in the atmosphere of Venus, managed to get a spectrum of Sirius B. Rather to his astonishment, he found that the surface temperature of Sirius B was 10,000°C (18,000°F), which was as hot as the surface of Sirius A and considerably hotter than the surface temperature of our sun.

This meant that every bit of Sirius B's surface was blazing out more light than equal portions of our sun's surface. Why, then, was Sirius B so much less luminous than the sun? The only answer was that there was very little surface to Sirius B; it was a dwarf, and a tiny dwarf, at that. It was the first to be discovered of a whole class of white-hot but very small stars now known as *white dwarfs*.

We now know that the diameter of Sirius B is only 11,100 kilometers (6,900 miles), so that it is smaller than Earth. But it must have the mass of the sun in order to exert enough gravitational pull to force Sirius A out of its path. How, then, can a mass equal to the sun be squeezed into a volume the size of a planet?

If we work out the density of Sirius B, it turns out to be about 33,000,000 grams per cubic centimeter, about 1,500,000 times as dense as the element osmium, the densest substance we know of on Earth. What's more, the surface gravity on Sirius B must be 462,000 times that on Earth.

A few years before Adams's discovery, these extreme figures would simply have been dismissed as ridiculous, for certainly nothing could be that dense. Even if you put osmium under enormous pressures, it was thought that you couldn't compress its atoms more than slightly. But just before Adams's discovery, Rutherford had shown that atoms consisted of a central nucleus, a very tiny one, that contained virtually all of an atom's mass. At the high temperatures and pressures in a star's core, the atoms break down and the atomic nuclei move about freely and compress much more tightly than they possibly could if the atoms were intact. Such broken-down atoms are called *degenerate matter*.

The sun has a core of degenerate matter, but a white dwarf is all degenerate matter. When a red giant collapses and becomes a white dwarf, the outer layers, still containing hydrogen, are blown off, leaving the star surrounded by a ball of gas that expands in all directions and finally disappears into outer space. For a period of time, however, newly formed white dwarfs appear to be surrounded by a doughnut of gas, for the edges of the ball absorb more light than the center does. What we see, then, is called a *planetary nebula*, because the gas looks as if it is filling a planetary orbit.

Once a white dwarf is formed, it expends energy so slowly that it lasts a long time before finally cooling down and dying. It is thought that no white dwarf has ever existed long enough to become dark, and that there are perhaps 3 billion white dwarfs among the stars in the universe, though they are so dim that we see only the ones that are fairly close to us.

86. WHAT IS A NOVA?

As we talk about stellar evolution and changes in the nature of individual stars, we move a long way from the old Aristotelian notion that the heavens are perfect and changeless. Yet stellar evolution is very slow, and merely watching the stars over the course of a lifetime, or even a few centuries, isn't going to show us much change.

Every once in a while, however, we do get an unmistakable impression of change, for a new star suddenly appears in the sky that had not been there before. The first recorded sighting of a new star in the sky was by Hipparchus, who is supposed to have seen one in 134 B.C. in the constellation Scorpio. We can't be sure of this, because the only record we have of the event was written by the Roman writer Pliny two centuries later.

After the decline of Greek astronomy in the 100s, the best astronomers in the world were the Chinese, who reported several new stars during the time between the 100s and the 1100s, all of which were particularly bright new stars. In 1006, they noticed a new star that was two hundred times as bright as Venus, and then in 1054, they sighted a new star that was two or three times as bright as Venus.

None of these stars were reported by European astronomers, partly because European astronomy was at a low ebb during this time, and partly because new stars, even very bright ones, are not easily recognized by people who are not constantly looking at the sky and memorizing the patterns of the constellations. Furthermore, European astronomers were so certain of the Aristotelian idea that the stars were changeless that even if they saw what they thought was a new star, they would probably have hesitated to report it.

All the new stars that were reported by the Chinese acted like stars in every way but one: They were not only new, they were temporary. They were bright points of light that did not move relative to the neighboring stars, so they couldn't be meteors or comets, and the brighter the new star appeared to be, the longer it lasted, but none of them lasted very long. Even the new star of 1006, which was so much brighter than Venus, could be seen in the sky for only three years, during which time it grew steadily dimmer until it gradually disappeared.

The turning point came in 1572, when a new star appeared in Cassiopeia. Again, it was several times brighter than Venus when it was first seen. It could even be seen by day, and on a dark, moonless night it cast a very faint shadow. By this time European astronomy was recovering, and the greatest astronomer of his time, Tycho Brahe, saw the star and studied it. He followed it every clear night for sixteen months, during which time it slowly faded and finally disappeared. He wrote a book about it, *De Nova Stella* (Concerning the New Star). As a result, such new stars have been called *novas* ever since.

Another new star, not quite as bright, appeared in the constellation Ophiuchus in 1604. It was observed and studied by Johannes Kepler.

Five years later, the telescope first came into use, and, little by little, astronomers invented all kinds of instruments with which to study the stars. Yet by a queer quirk of fate, no new stars have appeared in the sky since 1604 that were as bright as the brightest planets.

To be sure, novas appeared that were moderately bright, and in the 1800s, several novas were sighted that looked like first-magnitude stars but were not nearly as bright as Jupiter or Venus. In 1901, a nova named Nova Persei appeared in the constellation

Perseus. It was about as bright as Vega, and even brighter was Nova Aquilae, sighted in 1918, which was the brightest nova since 1604 and for a while was almost as bright as Sirius. Then there have also been Nova Herculis in 1934 and Nova Cygni in 1975.

Before the invention of the telescope, novas seemed completely strange to the people who watched them. They came out of nowhere and eventually faded back into space. Were they special messages from the gods designed to warn of disaster? Were they signs that the natural order of the heavens was breaking down? It was no wonder that none of the few astronomers of medieval Europe ever mentioned them.

The telescope changed everything, however. Nova Persei, for instance, didn't fade away and disappear. It merely became too dim to see with the unaided eye, though it remained visible through the telescope. This was true of the other novas of the 1900s, too. What's more, in photographs taken of the sky where novas later appeared, very dim stars were visible in just the right position.

What happens, apparently, is that a dim star suddenly brightens and for a short time increases its luminosity by hundreds of thousands of times, then fades, becoming the dim star it had been before. If careful photographs are taken of the star after it has passed through the nova stage, there are signs of a cloud of gas emerging from the star, which makes it appear as if it had gone through some sort of explosion and then returned back to its ordinary life.

But this merely raises another question; why should a star that has been shining steadily and quietly for an indefinite period suddenly explode?

In 1954, the American astronomer Merle F. Walker, studying the dim star that had been Nova Herculis twenty years before, found that it was a binary star—two stars circling a common center of gravity—and one of the stars was a white dwarf. This is the same situation that exists with Sirius A and Sirius B, but with an important difference: Sirius A and B are far apart and never approach each other more closely than about a billion kilometers, so that they revolve around each other in 50 years. The two stars of Nova Herculis, however, revolve about each other in 4½ hours, which means that they are very close together. In fact, they are separated by a distance of only 1.5 million kilometers (900,000 miles).

Thus, they have a powerful gravitational effect on each other, and hot hydrogen gas slowly leaks from the larger, normal star to the tiny white dwarf, with its terrifically intense surface gravity. If for some reason a larger than usual quantity of hydrogen leaked across to the white dwarf and settled down on its surface, the intense gravitational effect of the star would compress its material so tightly that it would undergo fusion in a flash. There would be a tremendous fusion explosion, and a nova would appear.

Since 1954, it has been found that all the medium-bright novas we can study are close binaries, with one of the component stars being a white dwarf. This means that we can feel certain that our sun will never suddenly become a nova, simply because it is not a binary star.

87. WHAT IS A SUPERNOVA?

The novas studied in the 1900s were by no means as bright as the monsters studied by Tycho Brahe and Kepler, or the earlier novas studied by the Chinese astronomers. In 1934, the Swiss astronomer Fritz Zwicky (1898–1974) gave the name *supernovas* to these very bright novas.

The study of supernovas (aside from just watching them and noting that they were very bright) began with the French astronomer Charles Messier (1730–1817). He was a comet-hunter who was occasionally fooled by a cloudy patch in the sky that proved not to be a comet. In the 1770s, therefore, he prepared a numbered list of the location of such patches in order to warn off other comet-hunters.

The objects on Messier's list are often known as M1, M2, and so on, after the numbers he gave them. They turned out to be of far greater importance than comets were. There is, for instance, the very first object on his list, M1, which is a foggy patch in the constellation Taurus.

M1 was studied in some detail in 1844 by the British astronomer William Parsons, the third Earl of Rosse (1800–1867). He had built a very large telescope for himself, which turned out to be useless because it was very difficult to maneuver and because the skies on his Irish estate, where he had the telescope, were hardly ever clear. However, he did study M1, and to him it looked like a turbulent cloud of gas within which there were crooked filaments of light. Because of those crooked filaments, he called M1 the Crab Nebula, and the name has stuck ever since.

It was studied again in 1921 by the American astronomer John Charles Duncan (1882–1967), who found that it was a little larger than Rosse had reported. The cloud appeared to be expanding, and the American astronomer Edwin Powell Hubble (1889–1953) suggested that, from its position, the Crab Nebula might well be the remains of the explosion that had created the supernova of 1054. The rate of the expansion was measured, and by calculating backward, it was determined that the original explosion had indeed taken place nine hundred years earlier.

A supernova, then, is the result of a stellar explosion, just as an ordinary nova, only the explosion is much greater. But what would cause the super-explosion?

The first hint at the answer came in 1931. In that year, the Indian astronomer Subrahmanyan Chandrasekhar (b. 1910), working in Great Britain, calculated how massive a white dwarf would be. The more massive it was, the more compressed it would be, due to the force of its own gravitational field, and Chandrasekhar calculated that past a certain point, it would simply break down. That point, which was called Chandrasekhar's limit, was reached when the star's mass was 1.44 times greater than the sun's. A white dwarf with a mass greater than that simply couldn't exist.

This limit did not seem very important at first, for at least 95 percent of the stars in existence are less than 1.44 times massive than the sun. These can all expand to red giants and then shrink to a white dwarf without any trouble.

Even very massive stars might be able to form white dwarfs, though, for when a massive star expands to a red giant and then collapses, only the inner portions collapse. The outer layers are left behind or blown outward so that a planetary nebula is formed. It seemed natural to suppose, then, that no matter how massive a red giant might be, the core that shrank would always be less than 1.44 times the mass of the sun and would form a white dwarf without trouble. (Actually, this turned out to be not quite so, as I shall soon explain.)

But now suppose you have a white dwarf that is almost 1.44 times the mass of the sun, but not quite, and suppose it is part of a close binary system, the other member being a normal star. The white dwarf keeps attracting matter from the normal star and adding the mass to its own. Even if the additional matter is hydrogen and undergoes fusion, it becomes helium that remains with the white dwarf. The result is that such white dwarfs slowly grow more massive, eventually gaining enough mass to push them past Chandrasekhar's limit.

When that happens, the white dwarf cannot maintain its structure and explodes. The explosion is millions of times greater than that of even the most remarkable ordinary nova. Such a supernova gleams with the light of a few billion ordinary stars for a while, then the light slowly dies down, and the entire white dwarf star is destroyed, leaving nothing behind. This explosion results in a *Type I supernova,* though there is also a *Type II supernova,* which is just a trifle less brilliant.

Clearly, our sun could never be a supernova. The white dwarf it formed would be well below Chandrasekhar's limit, and it has no binary companion from which it could gain mass.

The spectras of Type I supernovas show they have no hydrogen. This is to be expected if they are caused by exploding white dwarfs, because by the time a red giant collapses to a white dwarf, it has used up most of its hydrogen, and the central regions that undergo the collapse have none.

The spectra of Type II supernovas, however, show plenty of hydrogen, indicating that the explosion involves a star that has not yet reached the white-dwarf stage. It seems to be the red giant itself that explodes. The more massive a star is, the larger the red giant it forms and the more catastrophically it collapses. If it is large enough, the collapse is so sudden and drastic that all the hydrogen that is left in the collapsing portion is compressed, undergoes fusion, and produces a supernova.

The Type II supernova differs from the Type I in another way. Whereas the white dwarf that explodes as a Type I supernova leaves no trace of itself behind, the red giant that explodes and collapses as a Type II leaves a collapsed remnant.

However, this remnant does not become a white dwarf. For one thing, if the star is massive enough—say, at least 20 times greater than the mass of the sun—the remnant that collapses would be past Chandrasekhar's limit and would be too massive to become a white dwarf. Or the collapse could be so violent, with matter driven inward by gravitation with such force, that even if the mass of the collapsed portion were less than 1.44 times that of the sun, it would still compact itself beyond the white-dwarf stage.

But what happens if a collapsing star fragment moves beyond the white-dwarf stage?

In 1934, Zwicky and, independently of him, an American physicist, J. Robert Oppenheimer (1904–1967), both speculated on this question. They decided that a white dwarf must be composed of free atomic nuclei and free electrons, and that the electrons act as a kind of brake that prevents a collapse from proceeding too far. Yet the brake has only a limited capacity to halt condensation. If the mass is too large or the force of collapse too great, then the electrons are forced to combine with the protons in the free nuclei, forming neutrons. A star is then created that consists entirely of neutrons, which do not carry an electric charge and which come together till they touch one another. A star made up of only neutrons can squeeze all the mass of a star like our sun into a little ball not more than 14 kilometers (9 miles) across. This would be a *neutron star.*

It was an interesting speculation, but in the 1930s, there seemed no way that any object that tiny could be detected. If Sirius B were a neutron star instead of a white dwarf, it would still force Sirius A to adopt a wavy path, but it would shine with a light only $1/750,000$ as intense as it now does. It would have a magnitude of only about 20 or so, and would just barely be seen in our best telescopes. Sirius B, however, is the white dwarf closest to us. Astronomers thought that any other white dwarf, if it was actually a neutron star instead, would be entirely undetectable. So the whole notion went into limbo for over thirty years.

88. IS THERE ANYTHING USEFUL ABOUT SUPERNOVAS?

Astronomers believe that supernovas are vital, that without them, we wouldn't be here, life wouldn't exist on Earth, and the Earth itself would not exist. Consider the following: When the universe first formed, the only elements that came into existence were hydrogen and helium, the two simplest. (Of course, no one was present at the time to observe this, but scientists have worked out the possibilities—though not without controversy. The details are by no means certain, and I'll have more to say about them later.) The earliest stars consisted of hydrogen and helium, but conditions at the stars' centers made it possible for more complicated atoms to be formed there: carbon, oxygen, nitrogen, silicon, and even more complex elements such as iron. These more complex atoms stayed at the center of the stars, and even if a star became a red giant and then collapsed, these elements remained inside the condensed core.

It is only when supernova explosions occur that complex atoms are dispersed through space and are added to the clouds of gas in the universe, forming dust particles. When stars are formed from such "polluted clouds," we have a *second-generation star* that includes complex atoms to begin with.

Our sun is such a second-generation star. Every atom in the Earth and in our bodies (except for the occasional hydrogen atom) was once part of the interior of a star that later exploded. Without supernovas, our sun might be just hydrogen and helium, and Earth and life upon it wouldn't exist.

About 4.6 billion years ago, the solar system formed from a cloud of dust and gas containing complex atoms that had developed in the interior of stars and been strewn through space by supernova explosions. That cloud might, however, have already existed for billions of years. Why did it start contracting and condensing when it did?

We don't really know why, but one suggestion is that a nearby supernova sent out a blast that compressed a portion of the cloud nearest to itself. That intensified the gravitational pull in that part of the cloud and caused further contractions, which in turn brought the solar system, including the sun and Earth, into being. If this is true, then, once again, without supernovas, we would not be here.

Furthermore, biological evolution owes something to supernovas. When organisms duplicate themselves, they don't necessarily do so exactly, for if they did, then the earliest forms of life (simple bacteria-like organisms) would have never changed. As it is, with occasional imperfections, improvements come about more or less by accident, and very slowly, so that life-forms become more complex and better adapted to their surroundings.

There are various factors that bring about these imperfect duplications, but perhaps the most important and unavoidable are the cosmic rays (which we will have occasion to mention later on). These rays are produced by supernova explosions, and the fact that life on Earth has evolved beyond the bacterial stage is due to these explosions.

89. IS THERE LIFE ON PLANETS CIRCLING OTHER STARS?

Earlier, we came to the conclusion that there is probably no life of our type in the solar system outside Earth itself, though the satellites Europa and Titan might be extreme long shots. We might ask, then, if there is perhaps life on planets circling other stars.

Before we can really try to answer that, we have to ask if there *are* planets circling other stars. Over five hundred years ago, Nicholas of Cusa took it for granted that there were. Modern astronomers think he is likely to have been right, for if our solar system was formed from a cloud of dust and gas that automatically formed planets, that should be true of many other stars as well, and even, perhaps, of nearly all stars.

But that is risky reasoning. It would be much better if one star, aside from our own sun, were actually found to have a planetary system. Unfortunately, even with our present-day instruments, we can't see any planets circling other stars. Such a planet would be 4.4 light-years away, even if it were circling the very nearest star, and it would be shining only by the reflected light of that star, so that it would not deliver enough light to be seen at that distance. Even if it did, the much brighter light of the nearby star would drown it out. (The four large satellites of Jupiter are bright enough to be seen by the unaided eye, but the far greater light of nearby Jupiter drowns them out and we can see them only by telescope.)

the cluster is shaped like a globe, it is an example of what's known as a *globular cluster*. There are about a hundred globular clusters now known.

The globular clusters are distributed with an odd asymmetry. The British astronomer John Herschel (1792–1871) pointed out that they were not distributed evenly over the sky; almost all of them were located on one side of the sky, and virtually none on the other. In fact, fully one third of the globular clusters were in the constellation Sagittarius, which made up only 2 percent of the sky.

This observation, as we shall soon see, turned out to be of great importance.

91. WHAT ARE NEBULAS?

Not everything in the sky consists of stars or clusters of stars.

In 1694, Huygens saw and described a bright, fuzzy region in the constellation Orion. It looked like a luminous cloud and came to be called a *nebula*, which is simply the Latin word for *cloud*. The nebula described by Huygens is now known as the Orion nebula. We know it to be a huge cloud of dust and gas about thirty light-years across. If our entire solar system from the sun to the farthest comet were placed within the Orion nebula, it would be lost in the cloud's immensity, and even our sun and a dozen of its neighboring stars could all fit comfortably within it. In fact, the Orion nebula has many stars included within it, and it is those stars that cause it to shine by reflected light.

In 1864, the British astronomer William Huggins (1824–1910) managed to study the spectrum of the Orion nebula. It showed individual bright lines against a dark background, as one would expect of a hot gas, and the predication that it was a huge cloud (of perhaps the same type from which our solar system formed) was confirmed. Indeed, the Orion nebula is one place where astronomers are reasonably certain that new stars are being formed at this present time. Many other glowing nebulae of widely different shapes, some of remarkable beauty, have also been detected in the sky.

A nebula need not, however, necessarily be a glowing object. One that did not happen to have any stars within it would be a *dark nebula*. Herschel, for instance, spotted small dark areas in otherwise crowded starry regions, areas where no stars shine. He was puzzled by them and thought they might represent tunnels of starlessness, so to speak, the mouths of which just happened to be facing us. There were so many of them, however, that that explanation seemed inadequate. Surely there wouldn't be that many tunnels, all with their mouths directly facing us.

About 1900, E. E. Barnard and, independently, the German astronomer Max F. J. C. Wolf (1863–1932) suggested that the tunnels were dark nebulas that obscured the light of the stars behind them. Apparently, the starry skies were filled with clouds of dust that hid some of the glory of the universe, a fact that turned out to have great significance in our interpretation of how the stars are distributed in space.

---

92. WHAT IS THE GALAXY?

If we simply study the sky with our unaided eye, it would seem that we see stars in every part of it. There are no areas that seem particularly full of stars and none that are bare of them. From this we can conclude that the stars are distributed about us evenly in all directions, and if the stars as a whole form a collection that has a shape, that shape must be a sphere. That makes sense, too, since all sizable astronomical objects seem to be spherical; why shouldn't the universe as a whole be spherical?

Of course, what we see with our unaided eye are just 6,000 stars that are, by and large, rather close to us. What happens if we use the telescope? The answer is we see many, many more stars, but again, they seem to be distributed evenly around the sky—except for the Milky Way.

To the unaided eye, the Milky Way is a faint luminous band (which today we can hardly see if we live in cities, because the sky glows with human-made illumination). It has a faint milky appearance, and indeed, one myth tells us that once, when Hera, the wife of Zeus, was suckling an infant, some of her milk poured out into the sky and formed this faint luminous band. The Greeks called it *galaxias kyklos* (ring of milk), and the Romans called it *via lactea* (milky way), which is where we get the English name.

But what really is the Milky Way? If we dismiss mythology, we can begin with the Greek philosopher Democritus (c. 470–c. 380 B.C.), who suggested in about 440 B.C. that it was actually made up of a vast number of dim stars that could not be seen individually, but that all together gave rise to a soft luminosity. No one paid any attention to this view, but he happened to be perfectly right. This was proved when Galileo turned his first telescope on the sky in 1609 and found that the Milky Way contained myriads of stars.

How many are "myriads"? The first impression that people get when they look at the night sky is that the stars are innumerable, that they are too many to count. But as I have mentioned a few times already, the total number of stars visible to the unaided eye is only about 6,000. But through a telescope the number is much higher. Does that mean they are innumerable, after all?

The stars in the direction of the Milky Way are exceedingly numerous, but the stars in other directions are sparse in comparison, which means that we must abandon the notion of the entire body of stars forming a spherical structure. If that were so, the stars would be as numerous in all directions, as in the Milky Way, and the entire sky would be lit up luminously, with the nearer stars shining (less spectacularly than they are now) against that faintly luminous background.

We must suppose, then, that the stars exist in a large cluster that is *not* spherical in shape, but extends out in the direction of the Milky Way for far greater distances than in any other. It might be, then, that the Milky Way indicates that the stars are all clustered together in the shape of a lens, or a hamburger patty. This lens-shaped cluster is called the *Galaxy* (from the Greek expression for the Milky Way), while *Milky Way* is a name retained for the band of dim light we see encircling the sky.

The first person to suggest that the stars exist in a flattened galaxy was an English astronomer, Thomas Wright (1711–1786). He made the suggestion in 1750, but his ideas seemed so confused and mystical that few people paid any attention to him at first.

Of course, even if the Galaxy was lens-shaped, it might stretch out in the long diameter forever. There might be relatively few stars if you look away from the Milky Way, but an innumerable number in the direction of the Milky Way itself.

To settle the matter, William Herschel undertook to count the stars. Naturally, it was impractical to expect to count *all* the stars in any reasonable length of time. What Herschel did was to choose 683 small regions, well scattered over the sky, and count the stars visible in his telescope in each one. In this way, he took what we would now call a "straw poll" of the heavens. This was the first case of applying statistics to astronomy.

Herschel found that the number of stars in each region increased steadily as he approached the Milky Way in any direction.

must either be very massive or very close to us, and it is ordinarily impossible to tell which alternative is true. But since the Cepheid variables in the Small Magellanic Cloud are all considered to be about the same distance from us, distance can, in their case, be eliminated from consideration. If we note that one Cepheid is brighter than another in that cloud, we know that the brighter Cepheid is the more massive of the two and is actually the more luminous.

Leavitt found that in the Small Magellanic Cloud the brighter and more luminous the Cepheid, the longer its period of variation; there was a smooth relationship between the luminosity and the period.

Suppose, then, you found the distance of one particular Cepheid and measured its period. From these factors you would be able to determine its luminosity and derive the *luminosity-period curve* that Leavitt had discovered.

You could then study any other Cepheid in the cloud. From its period, you could find its luminosity by using Leavitt's curve, and from that, you would be able to tell how far away it must be to appear as bright as it does in the sky. In this way, the "Cepheid yardstick" could be used to measure distances of stars that are far too distant to show a measurable parallax.

One catch, though, was that even the nearest Cepheid was too far away to determine its distance by parallax, so we didn't have the distance figure that was necessary to set up the yardstick in the first place.

In 1913, however, Ejnar Hertzsprung (who discovered red giants) managed to work out, by means of a careful line of reasoning, the distance of some Cepheids without depending on parallax. That set up the yardstick.

In 1914, the American astronomer Harlow Shapley (1885–1972) applied the yardstick to Cepheid variables he located in various globular clusters. He found the distance to each and then designed a model of each cluster at their respective directions and distances. This gave him a three-dimensional model of all the globular clusters, which, he discovered, formed a more or less spherical ball, with its center thousands of light-years away in the direction of Sagittarius.

Shapley thought it reasonable to suppose that the sphere of globular clusters was at the center of the Galaxy, which thus seemed to be far away from us. He overestimated that distance, actually, and we now know that the sun is not located at or near the center of the Galaxy, but 30,000 light-years to one side.

In that case, why don't we see the Milky Way as far, far brighter in the direction of Sagittarius than it is in the opposite direction? Actually, the Milky Way is, to some extent, brighter and more complex in the Sagittarius direction than in any other, but we can't see the center of the Galaxy and beyond. The dark nebulas that litter the Milky Way obscure the large majority of stars in that direction.

What we see, then, when we look at the sky is only the rather small portion of the Galaxy that comprises the outer region nearest our solar system—our own neighborhood, in other words. If we consider only this portion of the galaxy, then, yes, we are near its center, but we are nevertheless far from its actual center.

94. WHAT IS THE DOPPLER EFFECT?

In order to find out more about the Galaxy, we must study another way of determining stars' motions. When Halley discovered that the stars moved, he could only measure the way they moved across the line of vision (the proper motion) as if they were sliding along the sphere of the sky. Once it became quite apparent, though, that the sphere of the sky did not exist and that stars were distributed nearer to us and farther from us through vast distances of space, the question arose: Is a particular star moving toward us or away from us? This motion, toward or away, is called *radial motion*, because the star is viewed as moving toward or away from us along the spoke (or radius) of a wheel stretching away from us, with the Earth at the hub.

How can we possibly detect this motion? If a star is moving directly away from us or toward us, its position in the sky would not change. Of course, if it moved away from us, it would appear dimmer and dimmer in the sky, and if it moved steadily toward us, it would appear brighter and brighter, but stars are so far away and move so slowly in comparison to their mighty distances, that it could easily take thousands of years for a star to change brightness enough to be detected even with delicate instruments. Furthermore, even if a star moved across the sky with a proper motion, it might also be moving toward or away from us so that it had a slanting motion in three dimensions. How could such motion possibly be seen?

The answer was found in a phenomenon observed on Earth that seemed to have nothing to do with stars. Thus, it was noticed that if a horseman was charging forward in a military attack, blowing a trumpet to hearten his troops and frighten the enemy, the trumpet seemed to change pitch as the horseman passed a stationary listener. At the moment of passing, the sound suddenly took on a lower pitch.

This phenomenon might have gone unnoticed in the heat of battle, but in 1815, the railroad locomotive was invented by the British engineer George Stephenson (1781–1848), and it was not many years before such locomotives were traveling as rapidly as a galloping horse, or even more rapidly. What's more, they usually had whistles of some sort to warn people when they passed through populated regions, so it became very common to hear the sudden lowering of pitch as the locomotive passed, and the question of why it happened arose.

The Austrian physicist Christian Johann Doppler (1803–1853) tackled the problem and decided, quite correctly, that when a locomotive was approaching, each successive sound wave caught up slightly with the one before it so that they hit the ear more frequently than they would have if the locomotive had been stationary. The pitch of the whistle was higher, therefore, than it would have been if the locomotive were stationary.

When the locomotive passed a listener and began to recede, each successive sound wave was pulled away from the one before it, and they hit the ear less frequently than if the locomotive had been stationery, so they sounded lower in pitch. As the locomotive passed, then, there was a natural change from higher than normal to lower than normal, from a high pitch to a low pitch.

In 1842, Doppler worked out a mathematical relationship between speed and pitch and tested it successfully by having a locomotive pull a flatcar back and forth at different speeds. On the flatcar were trumpeters, sounding various notes. On the ground, musicians with a sense of absolute pitch recorded the change as the train passed. Such changes in pitch were therefore called the *Doppler effect*.

By this time, it had been discovered that light also consisted of waves, though its waves were much, much tinier than those of sound. The French physicist Armand Hippolyte Fizeau (1819–1896) pointed out in 1848 that the Doppler effect should work for any wave motion, including light. As a result, the way it works with light is sometimes called the *Doppler-Fizeau effect*.

If a star is neither approaching nor receding from us, the dark lines in its spectrum should remain in place. If the star is moving away from us, the light it emits is longer in wavelength (the equivalent of deepening in pitch), and the dark lines all shift toward the red end of the spectrum (a *redshift*). The greater the shift, the more rapidly the star is receding from us.

If, on the other hand, the star is approaching us, the light it emits is shorter in wavelength (the equivalent of becoming higher in pitch), and the spectral lines shift toward the violet end of the spectrum. Again, the further the shift, the more rapidly the star is approaching.

If we know both the radial motion (in or out) and the proper motion (to one side), we can calculate the true motion of a star in three dimensions. But, actually, the radial velocity is the much more important of the two. The proper motion can be measured only if a star is close enough to make its motion across the sky rapid enough to be noticeable, and only a very tiny fraction of all the stars are that close to us. Radial motion, on the other hand, can be determined, no matter how far away a star is, provided its spectrum can be obtained.

In 1868, William Huggins was the first to determine the radial velocity of a star. He found that Sirius was moving away from us at about 46 kilometers (29 miles) per second. We have better figures now, but that was quite close for a first attempt.

95. IS THE GALAXY ROTATING?

Every object we know of in the solar system is rotating, from the sun to the asteroids, though some objects rotate more rapidly than others. This gives us the notion that other stars must also be rotating and that, in fact, even the entire Galaxy is rotating. But how can we tell if this is so?

Once astronomers were able to work out the true motion, in three dimensions, of a number of stars, it became possible to see that the stars were not moving in different directions randomly.

In 1904, for instance, the Dutch astronomer Jacobus Cornelius Kapteyn (1851–1922) found that a number of stars in the Big Dipper and elsewhere in the sky, too, were all moving in the same direction, more or less. He found, in fact, that there were two streams of stars, one moving with the stars of the Big Dipper and another moving in the opposite direction.

In 1927, J. H. Oort (who later advanced the theory of a distant cloud of comets) interpreted the existence of the two star streams in the following way. The stars of the Galaxy are all revolving about the galactic center. Those nearer the center than our sun are moving faster than the sun and are gaining on us, so that they all seem to be moving in one direction past us, with some, of course, moving more slowly than others. Stars farther from the center than the sun are moving more slowly, and the sun gains on them, so that they seem to be moving in the opposite direction from those that are nearer the center. So if all the stars are moving in the same direction, some faster, some slower, about the galactic center, it is fair to say that the Galaxy *is* rotating.

This rotation gives rise to an important conclusion. Astronomers have reason to believe that stars are closer together and exist in greater and greater thickness as we approach the center of the Galaxy. In fact, it seems likely that 90 percent of all the mass of the Galaxy is to be found in a comparatively small volume at the center. The stars outside the center circle that central mass, much as the planets of the solar system circle the sun.

Oort, in determining that the Galaxy rotated, showed that the sun revolved about the center of the Galaxy once in about 230 million years. From this period of revolution and from the distance of the sun from the center, it is possible to calculate the mass of the central conglomeration of stars.

It turns out that the mass of the Galaxy—or at least of the stars that make it up—is about 100 billion times that of the sun, which does not mean that there are 100 billion stars in the Galaxy, for the sun is not a representative star. Probably three quarters of the stars in the Galaxy are red dwarfs, and fully ninety percent are less massive than the sun. If the average star has half the mass of the sun, there might be approximately 200 billion stars in the Galaxy.

light-years away), the first indication that it had exploded
sudden appearance of seven detected neutrinos in a detect-
ice (a neutrino telescope) under the Alps mountain range.
cting devices improve, additional neutrinos may be found
tell us more about energetic events in the universe. In any
87 saw the beginning of "neutrino astronomy" outside the
stem.

other particle that comes to us from space is the *graviton*.
ns are uncharged, massless particles that, like neutrinos,
the speed of light. They are the least energetic particles
w of and the hardest to detect. Their existence was first
d by Albert Einstein in 1916, but all attempts to find them
ave failed, though physicists are sure they exist. If they
detected, they would probably yield information about
energetic events in the universe.

ht and light-like radiation travels through space as wave-
at exist in particle-like units. (All waves have particle
and all particles have wave aspects.) In 1905, Einstein
particle aspect of light a *photon*, from a Greek word for
far the greatest portion of the information we receive
universe even today is through photons, but not only
photons of visible light. There are photons of many other
th less and more energetic than those of visible light.

WHAT IS THE
CTROMAGNETIC
SPECTRUM?

When Einstein introduced the
concept of the photon, it became
clear that the shorter the waves of
a particular kind of light, the more
energetic the photons. Therefore
which has the longest waves in the spectrum, has the
getic photons. The photons of orange, yellow, green,
ight have successively more energy, and violet light has

96. DOES ANYTHING REACH US FROM THE STARS BESIDES LIGHT?

Until well into the 1900s, the only
information of consequence that
we received from the starry universe outside our solar system was
light. It was by means of light that
we studied the positions of stars,
their brightness, their motions, their temperatures, their chemical
composition, and even their gravitational effects on one another.
But there are other types of information reaching us from the stars.

After radioactivity was discovered in 1896, scientists learned
to detect radioactive radiations by means of various devices. Even
small quantities of such radiation were detectable. It was possible
to shield such detectors behind lead barriers that would absorb the
energetic radiations, and from the thickness of lead required to
block the radiations, the energy content could be estimated. What
surprised scientists was that when enough shielding was supplied
to block off all radioactive radiations they knew about, some particularly energetic radiation still managed to get through to the
detecting device. The question was, what was this unknown radiation?

It seemed to an Austrian-American physicist, Victor Franz
Hess (1883–1964), that whatever the radiation was, it had to come
from some earthly source. In 1911, he set about proving this by
carrying a detecting device, well shielded, high into the air in
balloons. He made ten balloon ascensions, five at night, going as
high as six miles. To his astonishment, he found that the higher
he went, the stronger the radiation he detected. At the greatest
heights he reached, the radiation was eight times as intense as it
was on Earth's surface, from which he could only conclude that
the radiation was not coming up from the Earth, but down from

space. Others began to investigate this radiation that was showering the Earth from all directions, and in 1925 the American physicist Robert Andrews Millikan (1868–1953) named the radiation *cosmic rays*, because they reached us from the cosmos.

The question remained as to the nature of cosmic rays. Millikan himself thought they were light-like waveforms that were much more energetic than light itself. Another American physicist, Arthur Holly Compton (1892–1962), thought, instead, that they consisted of very energetic electrically charged subatomic particles, traveling nearly at the speed of light. But how could this be proven?

If cosmic rays were light-like, they would be unaffected by Earth's magnetism; if they were charged particles, they would be bent by the field and more would reach the Earth near the magnetic poles than far away from them. In the 1930s, Compton traveled over the world, measuring the cosmic ray intensity here and there, and found that, indeed, it was more intense as one approached the magnetic poles. Cosmic rays were charged particles.

As it turned out, they were the same kind of charged particles emitted by the sun as solar wind. They were positively charged atomic nuclei, mostly hydrogen nuclei. When solar flares emit unusually energetic solar wind jets, the speeding atomic nuclei can be so energetic they are actually weak cosmic rays. The cosmic rays that reach us from beyond the solar system, however, are far more energetic than any that the sun can produce and are probably caused by events that are far more energetic than a mere solar flare, like a supernova explosion, for instance. Second, as cosmic rays pass among the stars, they are continually bent by the presence of magnetic fields, which tend to accelerate them and increase their energy.

Cosmic rays are important for many reasons. Earlier, I mentioned their influence on biological evolution, but they also tell us something about the chemical constitution of the universe generally and help physicists by producing collisions with atoms in the atmosphere that are far more energetic than any we can cause artificially even today. (However, waiting for cosmic ray colli-

sions of the proper sort can be very te
atom-smashing devices that can cre
milder, but can be had on demand

An important shortcoming of c
give us specific information about s
Their paths are so bent by magnetic
way of telling from what direction

Aside from cosmic rays, there
subatomic particles that have no n
through space at the speed of lig
predicted, on theoretical grounds, b
gang Pauli (1900–1958) in 1931.
neutrino (Italian for *little neutral o*

Since neutrinos have no mass
they interact with ordinary matter
most impossible to detect. Their e
physicists accepted their reality, bu
neutrino was actually detected by t
erick Reines (b. 1918) and Clyde

Unlike cosmic rays, they aren
by magnetic fields, but instead t
straight (except for the tiny effec
path of light. They are produced i
by other stars, and it is estimate
billion times as many neutrinos
subatomic particles.

The catch is that neutrinos p
liding with anything except at rar
only a few out of many trillions pa
For decades, for instance, Reine
neutrinos given off by the sun. F
detecting some, but only one thir
be expected. The reason for th
referred to as "the mystery of t

In 1987, when a supernova a
Cloud (the closest supernova to

150,000
was the
ing dev
As dete
that wil
case, 19
solar sy

An
Gravitc
travel a
we kno
predicte
so far h
could b
the mos

Lig
forms t
aspects,
called th
light. B
from th
through
types, b

97
ELEC

red light
least ene
and blue

the most energetic photons. The question is, though, whether photons of ordinary light are all the photons there are.

The answer is no, and that information has actually been known for nearly two hundred years, since 1800, when William Herschel discovered that the spectrum extended beyond the visible red end. You'll recall that he put a thermometer in different parts of the spectrum to see what temperatures he would obtain, and he found that the temperature was higher just beyond the red end of the spectrum than it was anywhere in the spectrum itself, which indicated some sort of invisible radiation beyond the red end of the spectrum. This light was called *infrared* (meaning *below the red*) radiation, and I have already mentioned it in connection with the greenhouse effect.

In 1801, the British physicist Thomas Young (1773–1829) finally demonstrated that light consisted of tiny waves rather than tiny particles. In 1850, the Italian physicist Macedonio Melloni (1798–1854) was able to show that infrared radiation had all the properties of ordinary light, except that its waves were longer and did not affect the eye. Once photons were understood, it was seen that infrared photons were individually less energetic than the photons of visible light.

Radiation also existed beyond the violet end of the spectrum. In 1801, the German physicist Johann Wilhelm Ritter (1776–1810) was testing the manner in which light caused the darkening of certain silver compounds. He found that this darkening was accelerated as he moved down the spectrum toward the violet end, but that it was faster than ever when he moved beyond the violet. Apparently, there was also *ultraviolet* (meaning *beyond the violet*) radiation, though it could not be seen, and we now know that ultraviolet light has shorter waves than visible light and that its photons are more energetic.

About 1870, James Clerk Maxwell produced four equations that described all the behavior of electricity and magnetism and showed that these two phenomena were different aspects of a single *electromagnetic interaction*. What's more, he showed that if the electromagnetic field vibrated, it produced a waveform that moved at the speed of light. If the vibration was of the proper speed, light itself was created, so that light could be considered an example of *electromagnetic radiation.*

At various speeds, however, the vibration could produce longer and longer waves, not only those of infrared radiation, but other forms of radiation far beyond it, and shorter and shorter waves, including those of ultraviolet radiation and beyond. In other words, there is an *electromagnetic spectrum,* with waves extending from the incredibly short to the incredibly long, of which visible light comprises only a tiny stretch.

Once Maxwell pointed out that such extreme radiations existed, scientists knew what to look for and were able to find them. In 1888, a German physicist, Heinrich Rudolf Hertz (1857–1894), discovered what came to be called *radio waves,* with wavelengths far longer than those of infrared radiation. In 1895, another German physicist, Wilhelm Conrad Roentgen (1845–1923), discovered X rays, which had wavelengths far shorter than ultraviolet radiation. Then in 1900, the French physicist Paul Ulrich Villard (1860–1934) discovered that among the radiations emitted by radioactive substances were *gamma rays,* an electromagnetic radiation with wavelengths shorter even than X rays.

Radio-wave photons were less energetic than infrared photons; X-ray photons were more energetic than ultraviolet photons; and gamma-ray photons were more energetic still.

Stars tend to radiate photons through the entire range of electromagnetic radiation. In that case, why is it we are sensitive to only the tiny range of the spectrum represented by visible light?

In the first place, a star like our sun delivers its peak intensity of radiation in the visible light region, so it makes sense for life-forms that depend on the sun to develop sensory equipment that receives and reacts to that range. Cooler stars, like red dwarfs, are much richer in the less energetic infrared radiation. Hotter stars, like the massive blue-whites, have many more of the energetic ultraviolet waves. Very energetic events in these hot stars can produce unusually high bursts of X rays and even gamma rays.

Second, Earth's atmosphere, while quite transparent to visible light, is relatively opaque to other portions of the electromagnetic spectrum, so we don't have much chance to become aware of other forms of light. But some of the infrared and ultraviolet waves near the visible spectrum do manage to pass through; for example, the ultraviolet light that penetrates the atmosphere is more energetic than visible light and much more effective in producing sunburn.

Beginning in the 1950s, human beings began to send rockets beyond the atmosphere, and equipped satellites placed into orbit around the Earth with devices capable of recording portions of the electromagnetic spectrum that could not penetrate Earth's atmosphere. By studying X rays emitted by the sun's corona, for instance, astronomers could prove that its temperature was a million degrees. By studying infrared radiations, astronomers found bands of dust about the bright star Vega that might indicate the existence of planetary bodies, and conducted a search for brown dwarfs. Ultraviolet emissions and even occasional bursts of gamma rays are also being studied.

There is no question, however, that in astronomy the most useful portion of the electromagnetic spectrum has come to be the region of the radio waves.

98. HOW DID RADIO ASTRONOMY DEVELOP?

In 1931, the American radio engineer Karl Guthe Jansky (1905–1950), working for Bell Telephone Laboratories, was tackling the problem of static that interfered with ship-to-shore telephone communications using radio waves. Static has a number of causes, including thunderstorms, nearby electric equipment, and aircraft passing overhead, all of which produce radio waves that interfere with the orderly radio waves used in telephone communication by introducing random effects that cause random sound—the crackling noise of static.

Jansky built a device that was capable of detecting these bothersome interfering radio waves, and while using it, he detected a new kind of weak static, a kind of hiss, whose source he could not identify at first. It came from overhead and moved steadily from day to day. At first it seemed to Jansky that it moved with the sun. However, it gained slightly on the sun to the extent of four minutes a day—just the amount by which the stars gain on the sun.

The source, Jansky therefore reasoned, must lie beyond the solar system, and by 1932 he had decided that it came from the direction of the constellation Sagittarius, the direction in which it was now known the center of the galaxy was located. Bell published his results before the end of 1932, and though they attracted little attention at the time, they marked the birth of *radio astronomy*.

One of the reasons Jansky's work drew so little notice was that astronomers of the time, never suspecting they might have to detect radio waves from the sky, lacked the proper equipment to receive and analyze such radiation. An American radio engineer, Grote Reber (b. 1911), tried to capitalize on Jansky's work, and in 1937, he built a radio-wave receiving device in the shape of a paraboloid (the shape of the casing that holds an automobile headlight). Thirty-one feet in diameter, it received and reflected radio waves, focusing them to the point where they could be more easily studied. In this way, he constructed the first *radio telescope* and became the first *radio astronomer*.

Reber discovered and mapped regions in the sky that seemed to be stronger-than-usual sources of radio waves. He called these *radio stars*, and the whole was a *radio map*. He published his results in 1942, but World War II was raging, and people paid little attention.

Nevertheless, the shortest radio waves, called *microwaves*, lying just next to the infrared portion of the spectrum, turned out to be very useful in the war. Such microwaves could be sent out in impulses that would be reflected by aircraft. From the direction in which the reflections came and from the time lapse between pulse and reflection, one could calculate where a plane was, in what direction it was flying, and how fast. This was called *radio detection and ranging*, where *ranging* means *distance determination*. The phrase was abbreviated to *radar*.

The development of radar moved quickly in Great Britain under the guidance of the British physicist Robert Alexander Watson-Watt (1892–1973). It was radar, more than anything else, that enabled the numerically inferior British Royal Air Force to defeat the German Luftwaffe in the Battle of Britain in late 1940.

In the process of developing radar, devices were invented that could detect radio waves, and once the war was over, it became possible to detect those coming from space with great precision. Larger and larger radio telescopes were built, for large radio telescopes could be built much more easily than equally large optical telescopes.

Of course, microwaves are much longer than waves of visible light, which meant one saw much more "fuzzily" by microwaves than by visible light. As radio telescopes became larger, however, microwave vision improved. Indeed, large radio telescopes could be built very far apart and could be synchronized by computers, so that the effect was of a radio telescope many miles in diameter. Microwave vision then became far sharper than light vision.

Since the 1950s, therefore, radio astronomy has been extremely useful in giving us information about the universe that ordinary light astronomy could not possibly have given us. As a result, we have learned more about the universe in the last thirty years than in all the time before.

99. WHAT ARE PULSARS?

Once astronomers zeroed in on microwave detection, they found it had two enormous advantages.

First, it was the only important region of the entire electromagnetic spectrum, other than visible light, to which Earth's atmosphere was transparent. There was a *microwave window* to the outer universe, as well as a *light window,* which meant that the microwave radiations of the universe could be studied from Earth's surface and it was not necessary to use rockets.

Second, microwaves could penetrate fog, mist, and dust clouds that were opaque to ordinary light. This was discovered in connection with radar during the war, for incoming planes could be followed even when they thought themselves hidden by fog or clouds. Similarly, portions of the outer universe that were opaque to visible light were transparent to microwaves, and we could study with microwaves what we could not see with light. Thus, the center of the Galaxy, forever hidden to our sight by dust clouds, could finally be studied through its microwave emissions.

Closer to home, it was the microwave emission of Venus, first detected in 1956, that gave astronomers their initial hint that the planet was extremely hot. What's more, rocket probes to Venus could send out microwave beams that penetrated the cloud layer and were reflected by the solid ground below. From these reflections, Venus' surface could be mapped, beginning in 1962, although it could never have been seen by visible light, except for the small patches that could be photographed by cameras dropped into the atmosphere.

Radar reflections could also be used to determine the rotation rates of Venus and Mercury. It was found that Venus rotated far more slowly than had been thought (and in the wrong direction), while Mercury rotated far more rapidly.

In 1955, the American astronomer Kenneth Linn Franklin (b. 1923) found a large emission of microwaves by Jupiter, which was finally explained in 1960, when it was shown that Jupiter had an enormous magnetic field, much larger than Earth's. This was confirmed in the 1970s, when probes were sent out beyond Jupiter.

More spectacular discoveries were made through radio astronomy in the universe beyond the solar system. As I mentioned earlier, Zwicky and Oppenheimer had independently speculated on the existence of neutron stars, extremely condensed stars consisting only of neutrons, which squeezed the mass of an ordinary star into a tiny ball just a few kilometers across. The possibility that such a neutron star had formed after the supernova explosion responsible for the Crab nebula was explored by the American astronomer Herbert Friedman (b. 1916). X rays were detected from various parts of the sky, and one source was the Crab nebula. Could the source be a neutron star remnant within the Crab nebula?

In July 1964, the moon was to pass in front of the Crab nebula, and Friedman supervised the sending of a rocket into space to monitor the X-ray production during the event. If the X rays were coming from a neutron star, then the X ray emission would be cut off entirely and at once as the moon passed in front of the tiny object. If the X-ray emission dropped off gradually as the moon moved in front of the Crab nebula, then the source was the entire nebula and not a tiny object within it. The latter proved to be the case, and those who had hoped to detect a neutron star in this way were disappointed.

PULSAR

In 1964, however, a new discovery was made. Radio waves from certain regions in the sky seemed to show a rapid fluctuation in intensity. It was as if there were "radio twinkles" here and there. The British astronomer Antony Hewish (b. 1924) designed a radio telescope that would make it possible to study fast changes in microwave intensities in greater detail. He supervised the construction of 2,048 separate receiving devices spread out over nearly three acres, and in July 1967, they were put to work.

Within a month, a young British graduate student, Jocelyn Bell, detected bursts of microwaves from a place midway between

Vega and Altair. The bursts were, as it happened, astonishingly brief, lasting only one thirtieth of a second. Even more astonishing, the bursts followed one another with remarkable regularity. They were so regular, in fact, that the period could be worked out to one hundred-millionth of a second; it was 1.33730109 seconds. By February 1968, when Hewish announced the discovery, he had located three other such radio twinkles, and since then hundreds more have been discovered.

Naturally, there was no way of telling, at first, what such a pulse represented. Hewish could only think of it as a pulsating star, with each pulsation giving out a burst of energy. This name was shortened almost at once to *pulsar*, by which the new objects came to be known.

All the pulsars are characterized by extremely regular radio pulses, but the exact period varies from pulsar to pulsar. One had a period as long as 3.7 seconds. In November 1968, astronomers detected a pulsar in the Crab nebula that had a period of only 0.033089 seconds; it was pulsing thirty times a second. Since then a few pulsars have been discovered that pulse several hundred times a second.

The question was, what can produce such short flashes with such fantastic regularity? Some object must be revolving, or rotating, or pulsing, and with each revolution, rotation, or pulsation, it must send out a microwave burst. To do this, however, it must revolve, rotate, or pulsate in a matter of seconds, or even hundredths of a second, which would require a very small size, combined with a very strong gravitational field. Pulsars can't be white dwarfs, for instance, because white dwarfs are too large and their gravitational fields are too weak. If they could be imagined as forced to revolve, rotate, or pulsate with sufficient speed, they would tear themselves apart.

The Austrian-American astronomer Thomas Gold (b. 1920) suggested almost at once that a pulsar had to be a rotating neutron star. A neutron star was small enough to rotate in a matter of a fraction of a second and would have a surface gravity intense enough to hold together when it did so. It had already been theorized that a neutron star would have an enormously intense magnetic field, with magnetic poles that need not be at the pole of rotation; electrons would be held so tightly by the gravitational pull of the neutron star that they would be able to escape only at the magnetic poles. As the electrons were thrown off, they would lose energy in the form of microwaves. If, as the neutron star rotates, the microwaves happen to be thrown off in our direction, we get one burst, or possibly two, for every rotation.

Gold pointed out that as the microwaves were given off, the neutron star would lose rotational energy and its period would very slowly increase. This hypothesis was tested for various pulsars and the increase was found. In particular, the period of the Crab nebula pulsar was slowing by 36.48 billionths of a second each day.

The Crab nebula thus had a neutron star inside it, after all. But other parts of the Crab nebula also emitted X rays. Only 5 percent of the X rays come from the pulsar, which was what misled Friedman. In 1969, astronomers found that the Crab nebula pulsar also emitted very brief flashes of light with each revolution. It flicked on and off 30 times a second, so it was called an *optical pulsar.*

The first really fast pulsar was located in 1982. It sent out radio pulses 642 times a second. It is probably smaller than most pulsars, perhaps not more than 5 kilometers (3 miles) in diameter and with a mass perhaps two or three times that of our sun. Other fast pulsars have also been located.

Sometimes a pulsar will suddenly speed up its period very slightly, then resume the slowing trend. Some astronomers suspect that such a glitch might be the result of a *starquake,* a shifting of mass distribution within the neutron star. Or it might be caused by a sizable body plunging into the neutron star and adding its own momentum to the star's.

100. WHAT ARE BLACK HOLES?

Back in about 1800, Laplace (who first advanced the nebular hypothesis) pointed out that the more massive and dense an object was, the higher its surface gravity and the higher its escape velocity. There were combinations of mass and density that would produce so high a surface gravity that the escape velocity would equal or surpass the speed of light. In that case, light could not be emitted by the object.

At the time, the hypothesis just seemed an idle speculation, for nothing was known that was nearly massive enough or dense enough (one or the other or both) to bring about such a situation.

In 1939, however, when Oppenheimer worked out the properties of a neutron star, he pointed out that if such a neutron star was more than 3.2 times the mass of the sun, even the neutrons of which it was composed would not be able to resist the gravitational pull inward. The neutrons would collapse, and there would be nothing else strong enough to resist gravity, which would therefore bring about a total collapse to a *singularity*, a point of virtually no volume and virtually infinite mass and density.

Such a *super neutron star* would not be able to give off light, as Laplace had predicted. It could be pictured as an infinitely deep "hole" in space, into which anything could fall, but nothing could get out. Since not even light could escape, the American physicist John Archibald Wheeler (b. 1911) suggested it be called a *black hole*, and the name stuck.

In 1970, however, the British physicist Stephen William Hawking (b. 1942) pointed out that black holes could very slowly "evaporate," so they were not absolutely permanent objects.

Black holes are most likely to form where stars are strewn most thickly, where stellar collisions are most frequent and stars might cling together to form huge masses that can collapse. It follows that black holes might be found at the center of globular clusters and are even more likely to occur at the center of the Galaxy.

Indeed, the small core of our galaxy in the direction of Sagittarius is so active—that is, it liberates so much energy (remember, Jansky's first discovery of radio sources outside our solar system was from this core)—that most astronomers are reasonably certain there is a black hole at the center, with a mass of perhaps a hundred million stars.

Such a massive black hole would continue to grow as it swallowed matter in the neighborhood, perhaps even entire stars, gulping them down whole, so to speak. There is no danger, however, of the entire galaxy being swallowed down in the near future. As a black hole clears the area around itself, there is less and less likelihood of further acquisitions.

The problem is, how does one actually observe a black hole to determine whether it really exists? Since it emits no photons of any kind, we can't see it anywhere along the electromagnetic spectrum. However, material trapped by the black hole's gravitational field revolves about it rapidly and, through collisions, loses energy and tends to spiral into the black hole. The process of spiraling produces X rays, so that wherever X rays are given off in the sky, we might at least suspect the possibility of the existence of a black hole. Unfortunately, there are other processes that might also release X rays, so that alone only raises the *possibility* of a black hole and nothing more. Thus, even if we know the center of the Galaxy is active and producing a great deal of radiation, that doesn't give us direct evidence of a black hole.

Suppose, though, a black hole is part of a close binary system, with a normal star as its companion. A close binary system consisting of a white dwarf and a normal star can give rise to novas. Close binaries of two neutron stars are also known to exist, and the study of their motion has been used to back up Einstein's general theory of relativity. Why not, then, a close binary system consisting of a black hole and a normal star?

If such a thing existed, matter would be drawn from the normal star into the black hole and would produce X rays as it spiraled down. Since matter would be drawn irregularly, the X rays would vary in quantity and intensity in an irregular fashion.

In 1965, a particularly intense X-ray source was detected in the constellation Cygnus and was named Cygnus X-1. In 1971, an X-ray-detecting rocket showed that the X rays emitted by Cygnus X-1 were irregular, indicating the possibility of a black hole.

Cygnus X-1 was immediately investigated with great care and was found to exist in the immediate neighborhood of a large, hot blue-white star estimated to be about thirty times as massive as our sun. This star and the X-ray source revolved about each other, and judging from the location of the center of gravity, the X-ray source seemed to be from five to eight times as massive as our sun. Since it could not be seen, it had to be a condensed star of tiny size, and since it was too massive to be a neutron star, it must be a black hole.

This is the nearest we've come to the actual detection of a black hole, and most astronomers accept Cygnus X-1 as such and are certain that black holes exist and might even be fairly common.

101. WHAT IS IN INTERSTELLAR DUST CLOUDS?

There are clouds of dust and gas in interstellar regions—the space between the stars. Scientists were initially quite sure that the dust consisted of fine grains of rocky materials and metals, the stuff that ended up making up the smaller worlds when such clouds condensed into stellar systems of stars and planets. As for the gas, that consisted mostly of hydrogen and helium.

Even though the dust and gas are thick enough to obscure the stars within and behind them, and plentiful enough to form stars and planets, this material is spread over such vast spaces that at first scientists were confident that the dust particles were tiny and the gases consisted of single atoms. They were just too widespread to stand much chance of striking one another and coalescing.

Information about the actual content of the clouds was first

obtained by the German astronomer Johannes Franz Hartmann (1865–1936) in 1904. He studied the radial velocity of the star Delta Orionis and found that the various spectral lines shifted by the same quantity, as was to be expected, though with some exceptions. The lines that represented the element calcium did not budge. It did not seem likely that the star would be moving and leaving calcium behind, and Hartmann felt that he was detecting calcium in the thin, largely motionless interstellar matter that lay between the star and ourselves.

Of course, the major component of the interstellar matter was hydrogen, and beginning in 1951, spectral lines representing ionized hydrogen (that is, hydrogen hot enough to lose the electrons from its atoms) were detected by the American astronomer William Wilson Morgan (b. 1906). The hydrogen was hot because of the near presence of large blue-white stars, which apparently existed in curved lines in the galaxy. The hot hydrogen marked out these lines, so that the structure of our galaxy could be viewed not as a simple lens shape, but more as a pinwheel with spiral arms extending from the central regions. Our solar system is in one of those arms.

Very little could be seen in the interstellar clouds if only visible light spectra were considered. With the coming of radio astronomy, everything changed, for cold atoms and atom combinations that emitted no light to speak of emitted far more of the less energetic microwaves.

In 1944, for instance, the Dutch astronomer Hendrik Christoffell Van de Hulst (b. 1918), in hiding during the German occupation of the Netherlands during World War II and unable to work on astronomy in the usual way, instead calculated how cold hydrogen atoms in space might behave. He realized that these hydrogen atoms could each line up their nucleus and their electron (they had only one electron apiece) either in the same direction or in opposite directions, and every once in a while, a hydrogen atom shifted from one configuration to the other, emitting a microwave with waves that were 21 centimeters (8.2 inches) long. Any given hydrogen atom would do this only every 11 million years or so, but there were so many hydrogen atoms in space that some were always producing those microwaves. In 1951, the American physicist Edward Mills Purcell (b. 1912) detected this microwave emission, and thereafter, it could be used to track unusual concentrations of cold hydrogen in interstellar space.

As methods for detecting microwaves improved, minor components of the gas clouds could be detected. For instance, there is a rare type of hydrogen atom in which the nucleus is twice as massive as in ordinary hydrogen atoms. Ordinary hydrogen is hydrogen 1, but the more massive type is *deuterium* (from a Greek word for *second*), or hydrogen 2. In 1966, microwaves were detected that were characteristic of hydrogen 2, and there were some indications that in the universe as a whole, 20 percent of all the hydrogen is in the form of hydrogen 2.

Atom combinations can be identified by their characteristic microwave emissions. For instance, next to hydrogen, the most common atoms in space that are capable of combining with other atoms are oxygen atoms. It is not surprising that once in a long time, an oxygen and a hydrogen atom might strike each other and cling together in a combination known as a hydroxyl group. Such a group would emit or absorb microwaves in four characteristic wavelengths, and two of them were observed in clouds in 1963.

Astronomers began to accept two-atom combinations in the thin interstellar matter, though combinations of three or more atoms still seemed improbable. Yet toward the end of 1968, the microwave fingerprints for water molecules (two hydrogen atoms and an oxygen atom, three altogether) and ammonia molecules (three hydrogen atoms and a nitrogen atom, four altogether) were also detected.

After that, numerous rather complex combinations were found, invariably containing one or more carbon atoms, and the science of *astrochemistry* was founded. Astronomers are still not quite certain how these complex molecules, some made up of as many as thirteen atoms, manage to be formed in the thin near-vacuum matter of space, but there is also the possibility that if we could send detecting devices into clouds of interstellar matter (which we can't so far, for they are many light-years away), we could detect still more complex groupings.

102. WHAT IS SETI?

Earlier, we speculated on the possibility of life on planets circling other stars. We have not yet mastered a technology that would make it possible for us to visit these planets, and aliens have not, to our knowledge, visited us. So it is much more practical to send a message than to send ships and intelligent beings in either direction. Messages do not involve the enormous expense of large rocket-powered starships, nor do they endanger anyone's life. Moreover, whereas starships might easily take centuries or millennia to reach even the nearest stars, messages can move at the speed of light (which, by the way, is the fastest possible speed, as Einstein showed in 1905) and therefore take only years or decades.

As it happens, our civilization is not sufficiently advanced to send a message by any means powerful enough to reach the distant stars with reasonable intensity. We can assume, however, that if there are alien intelligences out there, they might well be more advanced than we are, so that our role would be to try to detect the messages of others rather than to send out any of our own.

The question is, in what form are those messages likely to arrive? They aren't likely to be modulated cosmic rays, since these are wastefully energetic and follow curved paths, and therefore would be scattered and distorted without necessarily giving any indication of their point of origin. Neutrinos and gravitons would be too difficult to detect. That leaves us with photons.

As for photons, we first have to rule out beams of light, which wouldn't show up clearly amid the vast quantities of light produced by the stars, and any photons more energetic than those of visible light would be too wasteful. This leads us to assume that microwave photons are the most likely means by which messages can be sent.

The search for extraterrestrial intelligence, which is abbreviated to *SETI*, amounts, therefore, to carefully watching the sky for the existence of any radio signals that are not absolutely regular, as are pulsar bursts, or absolutely irregular, as are those arising from turbulent clouds. Neither of these types of signals would indicate the presence of an extraterrestrial intelligence. We would need signals that are irregular, but clearly not randomly irregular.

There have been searches of this kind since the 1960s that have been brief and limited and have detected nothing. What is really needed is an elaborate system of detection devices that can examine the entire sky in detail over a considerable period of time. Unfortunately, this would cost a great deal of money and labor, and thus far mankind, although willing to spend trillions of dollars a year on war and preparations for war, is not willing to spend far less for something like SETI. The feeling is that it is not likely to succeed and would just be money thrown away. There is even some suspicion, among those who have seen too many primitive motion pictures, that if we do anything to attract the attention of aliens, they might be encouraged to come to Earth and conquer us (as though any aliens could do more harm to us than we are busily engaged in doing to ourselves).

Actually, SETI could prove to be a profitable procedure, even if we found no messages. First, the attempts to put together devices that would be suitable for receiving such messages would undoubtedly succeed in improving our techniques for radio astronomy, and this would be useful in many directions, even if we ended up making no search.

Second, if we inspected the sky carefully and came across no messages, we would be sure to find many objects of interest that we would not have found without our new techniques and without a careful, dedicated search. Pulsars, for instance, were not found because anyone was looking for them. The discovery was serendipitous, an unexpected by-product of a scientific search.

Third, even if we detected some sort of message and could make nothing out of it (and it is very likely we would not be able to interpret the product of alien minds), the mere fact of its existence would prove that it is possible for intelligent beings to attain technological expertise well beyond our own without necessarily destroying themselves.

Fourth, if the aliens were interested in having us understand them, and if they deliberately made the message elementary enough for us to interpret, the way might be opened for us to learn a great deal and to advance our own knowledge far beyond the levels it might ordinarily reach in a given time.

103. IS THE GALAXY THE ENTIRE UNIVERSE?

Once Herschel had demonstrated that the stars formed a lens-shaped galaxy, it was assumed that *that* was the universe. And as its size came to be understood—100,000 light-years across and perhaps 2 billion stars—it certainly seemed large enough to represent a respectable universe. No astronomer before the 1910s had dreamed of a universe anywhere near that size.

And yet the Galaxy wasn't quite all there was to the universe. The Magellanic clouds, in which Leavitt had studied Cepheid variables and worked out the yardstick that could be used to demonstrate the true size of the Galaxy, lay outside the Galaxy. From their Cepheids, it could be shown that the Large Magellanic Cloud was 160,000 light-years away and the Small Magellanic Cloud was about 200,000 light-years away. Of course, the Magel-

lanic clouds could be looked on as satellites of the Galaxy, as the Moon is a satellite of the Earth and the planets are satellites of the sun. In other words, we can consider the Magellanic clouds as outlying districts, or suburbs, of the Galaxy.

Is there anything else that could possibly lie outside the Galaxy?

Suspicion, though not a very strong one, rested on the Andromeda nebula (which we've mentioned earlier in connection with Laplace's nebular hypothesis, for instance). The Andromeda nebula is visible as a small object of the fourth magnitude that looks like a faint, slightly fuzzy star to the unaided eye. The German astronomer Simon Marius (1573–1624) was the first, in 1612, to view it through a telescope. Messier included it in his list of fuzzy objects that were not comets. It was the thirty-first on his list, so the Andromeda nebula is sometimes called M31.

Laplace worked out his nebular hypothesis under the inspiration of the Andromeda nebula, which he thought looked like a mass of swirling gas, so that it might be a star and planetary system in the process of formation. Immanuel Kant, who in 1755 had preceded Laplace in thinking of such a hypothesis, had a different idea. He thought that objects like the Andromeda nebula were immensely distant systems of stars, and he called them *island universes*. He was quite right, as it happened, but his idea was ignored.

The Andromeda nebula did have a certain whirling look

about it, and between 1845 and 1850, Lord Rosse (the first to give the Crab nebula its name) observed over a dozen other nebulas that had the same whirling look. In fact, some looked like pinwheels or whirlpools. One object on Messier's list, M51, was so spectacularly whirlpool-like in appearance that it was called the Whirlpool nebula.

These whirling nebulas were called *spiral nebulas*, and the Andromeda nebula was one of them, but it was viewed so nearly edge-on that the spiral nature was not easily seen. By the year 1900, some 13,000 spiral nebulas had been sighted, and it might be argued that all of them were objects in the Galaxy, representing planetary systems in the process of formation. (Later, it was found that the Galaxy itself had a spiral structure, but this was not known in 1900.)

By that time, though, light spectra of astronomical objects were being studied. In 1864, William Huggins had taken the spectrum of the Orion nebula and showed it to consist of bright lines against a dark background, exactly what was to be expected from a mass of hot gas.

On the other hand, the spectrum of the Andromeda nebula, first obtained in 1899, showed the kind of spectrum one would expect from a star. Could it be, then, that the Andromeda nebula was a mass of stars, but so much farther away than the Milky Way or the Magellanic clouds that it was impossible to make out individual stars in it? If so, it must lie far beyond our galaxy, and so must other spiral nebulas, so that the universe might be enormously larger than our galaxy alone.

How could the matter be settled? If normal stars are just too far away to be seen in the Andromeda nebula (assuming it is indeed made up of stars), what about stars that are much brighter than ordinary ones? What about novas?

As a matter of fact, a nova did appear in the Andromeda nebula in 1885, and it was named S Andromedae. It became so bright that it was almost visible to the unaided eye. However, there was no telling whether it was really part of the Andromeda nebula or just a nova that had developed in the direction of the

Andromeda nebula and was shining in front of it, but had nothing to do with it.

The thing to do was to look for more novas, and an American astronomer, Heber Doust Curtis (1872–1942), did just that. By careful observation, he picked up the tiny sparkles of numerous novas in the Andromeda nebula. There were so many of these novas that there was no chance at all that they just happened to appear in different places in the direction of the nebula. No other equal-sized region of the sky produced so many novas in so brief a time. Therefore the novas that look as if they are in the Andromeda nebula really *are* there.

Second, most of the Andromeda novas were so faint they could barely be seen. They were much fainter than the novas that were indisputably a part of the Galaxy. The faintness of the Andromeda novas made it seem as if the nebula must be very far away—and far outside our galaxy. (Why, then, was S Andromedae so bright? Because, astronomers later decided, it had not been an ordinary nova, but a supernova.)

Curtis's notions, expressed in 1918, shook the astronomic world, which was reluctant to go along with him. In 1920, Curtis and Shapley (who had recently determined the size of the Galaxy) debated the matter, with Shapley sternly opposed to Curtis's notions. The debate ended in a kind of draw, but as time went on, it became more and more obvious that Curtis must be right.

The matter was finally settled by Hubble, who made use of a new hundred-inch telescope at Mt. Wilson Observatory in California. With it, he could make out individual stars in the outskirts of the nebula, which thus proved it to be a collection of stars and not just a mass of gas and dust. In 1923, he was able to identify one of the stars as a Cepheid and to use it to estimate the distance of the Andromeda nebula. His initial figure was too low, but it was high enough to show that the Andromeda nebula lies far outside our galaxy.

Nowadays, we know that the Andromeda nebula is 2.2 million light-years away from us, a distance equal to 22 times the total width of our galaxy. With that, the Andromeda became known as the *Andromeda Galaxy,* and the spiral nebulas as *spiral galaxies.*

So it seems that the universe consists of millions of galaxies, perhaps billions, and is enormously larger than the Galaxy alone.

104. ARE THE GALAXIES MOVING?

Since it was settled four hundred years ago that the Earth is moving about the sun, and one hundred and fifty years ago that the sun is moving about the center of the Galaxy, it should come as no surprise if we learned that the Galaxy is moving also.

A close study of the nearest galaxies shows that our own galaxy, the Milky Way, is part of a cluster of galaxies called the *local group.* The two chief members are the Milky Way and the Andromeda galaxy, the Andromeda being even larger than our own, with at least 300 billion stars. On the outskirts of the group is another large galaxy called the Maffei 1 galaxy (after the astronomer who first studied it), which may or may not be part of the local group. In addition, there are nearly twenty smaller galaxies, each containing 100 billion stars or less.

The galaxies within the local group, including our own, move majestically about the center of gravity of the entire system, and they can all be included in a sphere about 3.5 million light-years in diameter. Yet even so vast a sphere represents only our immediate neighborhood. Beyond it lie other clusters of galaxies, some of them much larger than the local group and containing thousands of galaxies.

We can presume that in every cluster of galaxies the individual galaxies move about some center of gravity, but how do the clusters themselves move?

The first hint of an answer to this question came even before astronomers realized that other galaxies existed in the universe. In

1912, the American astronomer Vesto Melvin Slipher (1875–1969) measured the radial velocity of the Andromeda nebula (as it was then called) and found that it was approaching us at 200 kilometers (125 miles) per second. Part of this approach is the result of the sun moving toward the Andromeda galaxy in the course of its own revolution about the center of our galaxy. If the Andromeda galaxy's approach to our galaxy is measured center to center, the speed is only 50 kilometers (30 miles) per second.

So far this seemed quite an ordinary state of affairs, but Slipher measured the radial velocity of fifteen nebulas altogether, and except for the Andromeda and one other galaxy (that eventually turned out to be part of the local group), all were moving away from us. What's more, some of the speeds of recession seemed unusually high.

Other astronomers took up the task, and they continued to find that *all* the galaxies (except for the two that Slipher had studied at the start) were receding from us. What's more, the dimmer the galaxies (and therefore, presumably, the more distant), the more quickly they were receding.

The American astronomer Milton La Salle Humason (1891–1972) took photographs that included exposures that continued for night after night in order to get the spectra of very faint galaxies. In 1928, he found a galaxy that was receding at a speed of 3,800 kilometers (2,350 miles) per second, and by 1936, he had one that was receding at 40,000 kilometers (25,000 miles) per second.

That sort of thing raised a problem. Why should all the galaxies be receding from us, and why should the recessions be faster and faster as they got farther and farther away? Was there something special about our galaxy? Did it repel other galaxies, and did this repulsion grow stronger with distance?

That did not make sense. If our galaxy exerted a repulsive force, that force should make itself felt within the local groups, and it didn't. What's more, a repulsive force that grew stronger with distance didn't seem likely. A magnetic pole can repel another magnetic pole like itself, and an electric charge can repel another electric charge like itself, but in each case, the repulsion weakened with increasing distance. There must be some other explanation.

Hubble, who had been the first to see actual stars within the Andromeda galaxy, tackled the question. He noted that the galaxies were not merely receding from us, they were receding from each other. No matter which galaxy we were to find ourselves on, it would seem to us that the all other galaxies were receding from us at a rate that increased with distance. Hubble concluded, in 1929, that the entire universe was steadily *expanding* and that the galaxies were moving apart from one another as part of the expansion and not because of any repulsive force.

As a matter of fact, Albert Einstein, in 1916, as part of his general theory of relativity, had prepared a set of equations that were intended to describe the properties of the universe as a whole. These showed that the universe would have to be expanding, though Einstein himself didn't realize this at the time.

105. IS THERE A CENTER OF THE UNIVERSE?

The sun is the center of the solar system, and all the planetary objects revolve about it. There is a central core to the Galaxy, and all the stars in the outskirts revolve about it. Is there then a center of the universe, a point from which all the galaxies are receding?

It would seem that there ought to be, but there isn't, because the expansion of the universe is not taking place in the usual three-dimensional fashion. It is four-dimensional, involving not only the three usual dimensions of ordinary space (length, width, and height), but also a fourth dimension of time. It is hard to picture a four-dimensional expansion, but we can explain it, perhaps, by an analogy with an expanding balloon.

Imagine the universe is a balloon that is expanding, and that the galaxies are spots on the surface, and that we live on one of those spots. Imagine that neither we nor the galaxies can ever leave the surface of the balloon. We can slide along it, but we can never move outward or inward. In a sense, we are picturing ourselves as two-dimensional beings.

If the universe continued to expand and the surface of the balloon stretched out and out, the spots on the surface would move farther and farther from one another. Anyone on one of the spots would see all the other spots receding, and the farther a particular spot was, the faster it would be receding.

Now imagine us searching for the place from which all the spots are receding. We would find it nowhere on the two-dimensional surface of the balloon. The true expansion is from the very center of the balloon, which is inward, in the third dimension, which we can't explore because we are confined to the surface.

In the same way, the place in the universe from which the expansion began is not anywhere in the three-dimensional space of the universe in which we can travel; it is someplace in time past, billions of years ago, and we can't travel there, although as we shall see, we can get information about it.

106. HOW OLD IS THE UNIVERSE?

If the universe is expanding, it was smaller yesterday than it is today, and smaller still last year. If we imagine ourselves moving further and further back in time, then the universe must once have been very small, indeed, and all the matter in it must have been compressed into a tiny volume.

The first to seriously consider such an idea was the Belgian astronomer Georges Edouard Lemaître (1894–1966). In 1927, he suggested that the universe began as a "cosmic egg," which exploded violently. The expanding universe of today would be the result of that explosion. The Russian-American astronomer George Gamow (1904–1968) called this explosion "the big bang," and the name has stuck.

But when did the big bang take place? If we know the average separation of the galaxies and the rate at which they are moving away from one another, it should be easy to calculate backward and see when they all came together.

There are several catches, though. In the first place, it is difficult to determine just how far particular galaxies are from one another. In the second place, it is difficult to tell just how rapidly they are flying apart. In the third place, it is not likely that the expansion has always been proceeding at the same rate.

When Hubble first decided that the universe was expanding, he made use of the best figures he could devise for average separation, rate of expansion, and change of expansion rate with time, and decided that the big bang had taken place 2 billion years ago. This estimate was met with a roar of disapproval from geologists and biologists, who were quite convinced that the Earth was much older than 2 billion years, and who insisted that the universe could not very well be younger than the Earth.

In the sixty years since Hubble's first estimate, additional information has set the big bang further off into the past. The figure commonly used now is that the big bang took place 15 billion years ago, and that the universe is therefore 15 billion years old. This is not hard and fast. There are some astronomers who argue for an age of 10 billion years, and some for 20 billion. Presumably, as more and better evidence is gathered, some decision on the matter will be reached.

If the 15-billion-year figure is correct, then at the time our solar system was formed, the universe had already existed for 10 billion years.

107. WHAT ARE QUASARS?

I have said that we can't travel into the past in order to check on when the big bang took place or what the circumstances were—but we can *see* into the past.

Whenever we look at a distant object, we know that the light we see it by (or the radio waves) has taken a certain length of time to reach us. Radiation cannot travel faster than light (roughly 299,800 kilometers or 186,300 miles per second) under any circumstances, and we see the object only as it was when the radiation started out on its journey, not when it finished it. Thus, when we look at the Andromeda galaxy, we must remember that the light by which we see it left the galaxy 2.2 million years ago, so we see it as it was 2.2 million years ago.

Of course, the Andromeda galaxy probably looks very much the same now as it did 2.2 million years ago, so in this case, the delay doesn't mean much. But what if we look at objects that are much farther away? What are the farthest objects we can see?

These farthest objects were seen before we had any idea they were particularly far away. As radio telescopes improved and the picture we saw by microwaves sharpened, it became possible to narrow down certain radio sources to very small regions. These were *compact radio sources,* and among them were objects known as 3C48, 3C147, 3C196, 3C273, and 3C288. The *3C* is short for *Third Cambridge Catalog of Radio Stars,* a listing compiled by British astronomer Martin Ryle (1918–1984).

In 1960, the American astronomer Allan Rex Sandage (b. 1926) investigated these sources and found that all of them seemed to arise from dim stars of the sixteenth magnitude that appeared to be part of our galaxy. This was very unusual, for individual stars were not generally sources of detectable microwaves. We receive them from the sun because it is so close to us, but not from other stars, not even those that are only a few light-years away. Why, then, should microwaves be received from these dim stars? Astronomers felt that they might not be normal stars, and so were called *quasi-stellar* (that is, *starlike*) radio sources. In 1964, the Chinese-American physicist Hong-Yee Chiu abbreviated *quasi-stellar* to *quasar,* and the name stuck.

But what were the quasars? In 1963, the Dutch-American astronomer Maarten Schmidt (b. 1929) puzzled over the spectrum of 3C273. The lines seemed altogether strange until it suddenly struck him that they were familiar lines that ordinarily occurred far in the ultraviolet; they had merely been redshifted to an enormous degree, which was why he had not recognized them.

From the redshift, it could be calculated that 3C273 was not an ordinary star of the Galaxy but an object that was about a billion light-years away, farther away than any ordinary galaxy that had yet been detected. The other quasars are even farther away—3C273 is the nearest of the quasars. Hundreds of them are now known, and some of them are 10 or 12 billion light-years away.

Now the problem is, how can these objects be seen at such distances? We must assume that they are more luminous than galaxies, that they are as bright as a trillion suns and up to 100 times brighter than an ordinary galaxy.

At the same time, it was found that the radiation they emitted was variable, and sometimes there was considerable variation in a matter of just a few weeks. This would indicate that the quasar could not be more than a few light-weeks (say, a trillion kilometers or so in diameter), since otherwise whatever influence caused the variation could not pass from end to end in a short enough time, because *nothing* can be transmitted faster than light. How can an object so small deliver so much energy?

The most likely answer dates back to 1943, when the American astronomer Carl Seyfert observed a galaxy that had a very bright and a very small nucleus. Other galaxies of this sort have been observed, and the entire group is now referred to as *Seyfert galaxies.*

The nuclei of Seyfert galaxies are very active, possibly because they contain unusually large black holes that are wreaking havoc on their cores. Perhaps the quasars are particularly large and bright Seyfert galaxies, and all we see, at their great distance, are their tiny, extremely active and luminous cores. Indeed, recent studies have shown that the quasars are surrounded by fuzziness that might represent the outer regions of a galaxy.

Because the quasars are, for the most part, located at enormous distances of billions of light-years from us, they must have flourished billions of years ago in the youth of the universe. Perhaps when the galaxies were young, a large number of them collapsed catastrophically at their centers into black holes. With time, those black holes absorbed all it was easy to absorb, and the galaxies settled down into objects that were quieter and more staid, so that all quasars "cooled" out of existence by a billion years ago.

This alone would show that the universe in its youth was quite different from what it is now and that there has been an evolutionary process. This tends to disprove competing theories that would have the universe possess no true beginning and that describe it as having had the same overall appearance at all times in the indefinite past.

108. CAN WE SEE THE BIG BANG?

No matter how far into the distance we penetrate, we cannot see the big bang itself. In recent years, it has been reported that galaxies have been seen at a distance of perhaps as much as 17 billion light-years (which would seem to indicate that the universe is at least 17 billion years old) and that they are so numerous that they seem to lie on top of each other—which is not surprising, for of course, the universe was far smaller 17 billion years ago than it is now, and galaxies must have been much closer together.

Yet we still can't see the big bang itself, at least not by light. In the early days of the universe, space was not transparent as it is today, but was filled with a fog of energy. Wherever we look, therefore, we are likely to end up with a fogginess we cannot penetrate.

That, however, only concerns light. In 1949, Gamow, who invented the term *big bang,* pointed out that we should still be able to sense a faint, far-off echo of the big bang. As a result of that cosmic explosion, there should be microwaves reaching us from the big bang, which can penetrate the fog. He even predicted the exact energy content of those microwaves.

Furthermore, as telescopes look farther and farther into the distance and, therefore, farther and farther into the past, they tend to follow a line that spirals inward as the universe shrinks steadily with passage backward in time. In whatever direction we look, the spiral leads us to the center and to the big bang. Therefore, Gamow predicted, the microwaves would come from all parts of the sky equally and with the same energy and characteristics everywhere.

In 1964, the German-American physicist Arno Allan Penzias (b. 1933) and an American physicist, Robert Woodrow Wilson (b. 1936), detected this uniform microwave background at just about the energies that Gamow had predicted. This is taken as the best evidence yet that the big bang really took place.

Astronomers are now trying to work out the events that happened in the early moments of the big bang. They reason that if they look backward in time, they can see the objects of the universe coming together and crashing into one another, so to speak, like a movie being projected in reverse. The result must be the same as when the material of the solar system came together to form the sun and the planets. The temperature rose and created a hot center of the Earth and a hotter center of the sun. If we look back in time at *all* the matter of the universe coming together, it should form a far hotter center of the universe. In other words, at the beginning, the universe was very tiny and incredibly hot, and it has been expanding and cooling ever since.

Scientists, allowing for the incredibly high temperatures, have advanced speculations as to the course of events in the first fractions of a second after the big bang. They are puzzling over

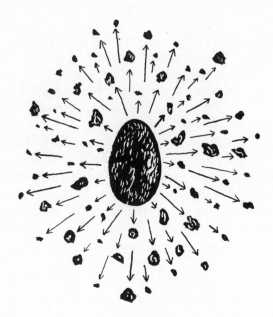

what happened in the first millionth of a trillionth of a trillionth of a trillionth of a second. These are just speculations right now, and we might as well let them go until enough evidence has accumulated to lend more substance to them.

109. HOW DID THE BIG BANG ARISE?

Until quite recently in time, most people in the West thought the Earth and the sky was formed by supernatural creation about six thousand years ago. (Many people today still earnestly believe this, though their intellectual achievement in doing so is about on a par with those who still believe the Earth is flat.) Today, however, scientists generally accept as fact that the solar system was formed by natural processes from a cloud of dust and gas 4.6 billion years ago, and that the cloud had existed since soon after the origin of the universe, perhaps 15 billion years ago.

But even if we go back to the big bang and imagine that all the matter and energy of the universe was concentrated into a tiny ball of incredibly dense, incredibly hot material that exploded to form the universe, where did that tiny ball come from? How did it come into existence? Must we assume supernatural creation at that stage?

Not necessarily. A branch of science was worked out in the 1920s called *quantum mechanics,* which is far too intricate to go into here. It has been an extremely successful theory, explaining phenomena nothing else could adequately explain and predicting new phenomena that turned out to behave precisely in accordance with the predictions.

In 1980, an American physicist, Alan Guth, took up the problem of the origin of the big bang in terms of quantum mechanics. We might visualize the universe before the big bang took place as a vast, illimitable sea of nothingness. Apparently, though, that is not an accurate description. The nothingness contains energy, and it is not quite a vacuum because, by definition, a vacuum contains nothing at all. The pre-universe had energy, and since all of its other properties resemble those of a vacuum, it is called a *false vacuum.*

From this false vacuum, a tiny point of existence appears where the energy just happened, by the blind forces of random changes, to have concentrated itself. In fact, we might imagine the illimitable false vacuum to be a frothing, bubbling mass, producing bits of existence here and there as an ocean wave produces foam. Some of these bits of existence might disappear promptly, subsiding back into the false vacuum. Some, on the other hand, might be large enough, or have been formed under such conditions, as to undergo a rapid expansion into a universe: We live in such a successful bubble.

This model has many problems, however, and scientists are

still struggling to patch it up and solve them. When and if they do, will we have a better notion of where the universe came from?

Of course, even if some version of Guth's theory is right, we might simply take a further step backward and ask where the energy of the false vacuum came from in the first place. That we can't say, but it won't help us to suppose that at that point there was a supernatural creation, for then we could take a further step backward and ask where the supernatural entity came from. The answer to such a question is usually a shocked "He didn't come from anywhere. He always *was.*" That's a hard thing to visualize, and we might just as well say that the energy of the false vacuum always was.

110. WILL THE EXPANSION OF THE UNIVERSE CONTINUE FOREVER?

Is there anything that acts to slow down and stop the the universe from expanding?

The only force we know that would do so is the mutual attraction of all parts of the universe—the gravitational force. The universe is expanding against its own gravitational pull, so the process of expansion must spend energy to overcome the pull. In doing so, the expansion slows and it might eventually stop. In that case, the universe, after a short pause, will begin to fall together again until it ends in a big crunch, which is, of course, the opposite of the big bang. If the universe expands forever, it is called an *open universe,* and if it eventually stops expanding and begins to contract, it is called a *closed universe.*

The same problem would face us if we considered an object thrown upward from the Earth's surface against the pull of gravity. It is our common experience that such an object, sent upward under ordinary circumstances, is eventually defeated by the gravitational pull of Earth. Its speed of rising decreases to zero, and then it begins to fall back to Earth. The more forcefully it is hurled upward and the greater, therefore, its initial upward speed, the higher it climbs and the longer it takes to begin to fall back.

However, the gravitational pull of the Earth weakens with distance. If an object is fired upward with sufficient speed, it rises so high that the weakening pull of the Earth will not suffice to slow its motion altogether. It continues to outpace the pull of Earth's gravity and never returns. The speed with which an object must begin its upward motion must be more than 11 kilometers (7 miles) per second if this is to happen. That speed, the escape velocity, is what rockets that are sent to the moon and beyond must generally attain.

We might ask, then, if the rate of the universe's expansion outward against gravitation's inward pull has achieved escape velocity. In order to come to a decision, scientists must estimate the rate of expansion. They must also estimate the average density of matter in the universe, for that would give them an idea of the strength of the inward gravitational pull. Both determinations, the rate of expansion and the average density of the universe, are hard to carry through, and the results are only approximate.

The conclusion is, however, that the actual density of matter in the universe is only about 1 percent or so of that required to end the expansion. The universe would therefore seem to be open, and to be expanding forever. That is so, however, only if we count the matter that we can see. If there is additional matter that we can't see or sense in any way, then the universe might be closed after all.

111. IS THERE MATTER IN THE UNIVERSE WE CAN'T SEE?

Astronomers think there must be. On several occasions, I have pointed out that gravitational influence tells us things that light does not. Thus, Sirius B was discovered by its gravitational influence on Sirius A before it was ever seen. The planet Neptune was discovered by its gravitational influence on Uranus before it was ever seen, and so on.

In galaxies, all mass appears to be concentrated toward the center, and the stars on the outskirts circle the galactic core much as planets circle a star. Therefore we would expect the stars to be circling the centers of their galaxies more and more slowly the further out they are from the core. This is true of our solar system, where the planets move more slowly with increasing distance from the sun.

We can determine the speed of a galaxy's rotation by measuring its radial velocity at different distances from the core. It turns out that, in galaxies in which such measurements can be made, the stars move about the galaxy's core at about the same speed, regardless of how far they are from it.

This observation defies the law of gravity, and scientists do not want to abandon that law. As an alternative, they must suppose that the mass of galaxies is *not* concentrated in the core, but is spread out much more evenly through the galaxy. Yet how can that be true when we *see* that the mass, in the form of stars, *is* concentrated toward the center?

Another mystery is that the galaxies in a given cluster tend to hang together, held in the grip of mutual gravity. However, if we calculate the pull that each galaxy ought to have from the stars it contains, and the speed with which the galaxies in a cluster are moving relative to each other, we come to the conclusion that there isn't enough gravitational pull to hold the clusters together. Yet they *do* hold together, which can only mean that there is additional matter in the clusters that we cannot see, but that is present in sufficient quantity to supply the necessary gravitational pull to keep the clusters in place.

What can this additional invisible matter be? Astronomers do not yet know the answer, and they refer to it as "the mystery of the missing matter." There are many speculations, but we must await the arrival of further evidence before we can be reasonably certain as to what this matter is or whether it exists at all. We can say that if the missing matter *does* exist, there might be a sufficient quantity of it to close the universe and assure us that someday, perhaps trillions of years hence, everything will start contracting again. This is just one example, despite all our successes and all our accomplishments, of the many puzzles that still face us in the world about us.

INDEX

ABOUT THE AUTHOR

Isaac Asimov has published over 460 books of science fiction and nonfiction covering every branch of science and mathematics, history, literature, humor, and various miscellaneous subjects. He lives with his wife, Janet, in New York City.